THE EFFECTS OF UNDERNUTRITION ON CHILDREN'S BEHAVIOR

Food and Nutrition in History and Anthropology
A series edited by Solomon H. Katz, University of Pennsylvania

Volume 1 **FOOD, ECOLOGY AND CULTURE: Readings in the Anthropology of Dietary Practices**
Edited by John R.K. Robson

Volume 2 **FAMINE: ITS CAUSES, EFFECTS AND MANAGEMENT**
Edited by John R.K. Robson

Volume 3 **INFANT CARE AND FEEDING IN THE SOUTH PACIFIC**
Edited by Leslie B. Marshall

Volume 4 **FOOD ENERGY IN TROPICAL ECOSYSTEMS**
Edited by Dorothy J. Cattle and Karl H. Schwerin

Volume 5 **THE INFANT-FEEDING TRIAD: Infant, Mother, and Household**
Barry M. Popkin, Tamar Lasky, Judith Litvin, Deborah Spicer and Monica E. Yamamoto

Volume 6 **THE EFFECTS OF UNDERNUTRITION ON CHILDREN'S BEHAVIOR**
David E. Barrett and Deborah A. Frank

Additional volumes in preparation

This book is part of a series. The publisher will accept continuation orders which may be cancelled at any time and which provide for automatic billing and shipping of each title in the series upon publication. Please write for details.

THE EFFECTS OF UNDERNUTRITION ON CHILDREN'S BEHAVIOR

DAVID E. BARRETT
Clemson University

DEBORAH A. FRANK
Boston University School of Medicine
Boston City Hospital

Gordon and Breach Science Publishers
New York London Paris Montreux Tokyo Melbourne

Gordon and Breach Science Publishers

Post Office Box 786
Cooper Station
New York, New York 10276
United States of America

Post Office Box 197
London WC2E 9PX
England

58, rue Lhomond
75005 Paris
France

14–9 Okubo 3-chome
Shinjuku-ku, Tokyo 160
Japan

Camberwell Business Center
Private Bag 30
Camberwell, Victoria 3124
Australia

Library of Congress Cataloging-in-Publication Data

Barrett, David E., 1947–
The effects of undernutrition on children's behavior.

(Food and nutrition in history and anthropology, ISSN 0275–5769 ; v. 6)
Bibliography: p.
Includes index.
1. Malnutrition in children—Complications and sequelae. 2. Malnutrition in children—Psychological aspects. 3. Human behavior—Nutritional aspects. 4. Child development. I. Frank, Deborah A. II. Title. III. Series.
RJ399.M26B37 1987 155.4 87–6916
ISBN 2-88124-190-5

Printed in the United States of America.

CONTENTS

FOREWORD

To borrow a Czech turn of the phrase, Forewords are rarely, if ever, written "with a cool brain." At their best they are contributed by individuals who care—about the author or authors, the work itself, and the potential readers. I am ready to admit, without the application of rack and wheel, that I have been fascinated by the very idea of this book, that I have followed with a genuine interest its progress, and that I am impressed by the final product.

The book has much to offer to mature scholars but it will be particularly useful to graduate students and to young investigators who are just beginning to explore the area of malnutrition and behavior (cf. Brožek and Schürch 1984, Brožek 1985).

I wish such a treatise had been available fifty years ago when I was first struck by the potential magnitude of the impact of severe malnutrition on man's psychological functioning, even though my "independent variable" was iodine deficiency, not generalized malnutrition—the topic of the present work.

The volume provides a comprehensive account of the effects of energy-protein deficits on the behavior of children, opening with considerations of the definition, assessment, and prevalence of this type of malnutrition, with focus on undernutrition. A large chapter is devoted to general methodological issues: the central concept of "validity"; the varieties of research design applicable to the study of the relationships between nutrition and behavior; methods for the assessment of the behavioral variables, with special attention to "reliability"; and the relationship between theory, research design, and the operationalization of the behavioral variables. An extensive review of the literature takes into account both the animal and the human studies. The volume closes with a detailed presentation of an empirical study on the social-emotional functioning of children in the context of undernutrition and nutritional supplementation.

If pressed to identify the cardinal virtue of this book, I would submit that the tail—the account of the empirical study—wags the dog. The demonstration that chronic, moderate energy-protein malnutrition, experienced in childhood, may adversely affect the child's personality at an early school age is an important factual contribution to the available body of information on malnutrition and behavior. At the same time, the significance of the innovative methodology goes substantially beyond the realm of psycho-nutritional research: it may be expected that the procedures developed for the characterization of social and emotional aspects of child behavior may provide indicators that will prove to be sensitive to other biological stresses and to environmental disadvantages.

While stressing the significance of the methodology, in the context of nutritional research and beyond it, and of the results, we can not neglect the service rendered to young investigators by the book's emphasis on the need for the integration of theory, research design (with its implications for the statistical analysis of the data), and the outcome of the research. The data interpretation feeds back into the theory and has implications for future research as well as for practice; in the broad area of public health, behavior is an outcome of major importance (cf. Barrett 1986).

The content of the book is specified in detail in the Table of Contents, and the rationale of the individual parts of the book is spelled out in the Preface. Here I shall limit myself to providing information on the historical background of the book, with focus on the empirical study. It happens that I have interacted, at one time or another, with all the *dramatis personae* and that, I believe, these contacts influenced positively the course of the events.

Let us begin with a bit of prehistory. Who or what was at the beginning of the story? Importantly, a graduate seminar on child psychology, given by Professor Florence Goodenough in the year 1940/41 at the University of Minnesota. It was at this seminar that I met Marian Radke, better known as M.R. Yarrow, the moving spirit behind the empirical study which constitutes the cornerstone of the present volume. As Director of the Laboratory of Developmental Psychology, a research unit within the National Institutes of Mental Health, housed in building 15 K of the "campus" of the National Institutes of Health at Bethesda, Maryland, Dr. Radke-Yarrow introduced Dr. D.E. Barrett to the area of research on malnutrition and behavior. I do not know for certain but it is not only possible but probable that she was aware of the investigations on the effects of malnutrition, carried out in the early 1940s in the Laboratory of Physiological Hygiene, School of Public Health, University of Minnesota, culminating in the collaborative treatise on *The Biology of Human Starvation* (Keys et al. 1950). This may have sensitized her to the topic.

Be that as it may, in the summer of 1976, when Dr. Barrett was a Staff Fellow of the National Institute of Mental Health, Dr. Radke-Yarrow asked him if he would be interested to take part in a study on the effects of malnutrition on children's social behavior. As it happened, there was a guest research worker at the Laboratory of Developmental Psychology, Dr. Prebusha Ramalingaswami, who was exploring the possibilities of carrying out, in India, a naturalistic observational study on psychological-social effects of malnutrition, and was seeking professional assistance in designing and carrying out the study. While Dr. Barrett had no previous contact with nutritional research, he was trained in research design and had been engaged in investigations on prosocial behavior (Yarrow, Waxler et al. 1976; Barrett & Yarrow 1977; Barrett 1979). He found the Indian project interesting and challenging. He began to read, voraciously, the literature on the behavioral effects of energy-protein malnutrition and, in cooperation with Dr. Radke-Yarrow, immersed himself in the design of the Indian study.

At this stage of the developments I was asked, in October 1976, to give a seminar at the NIMH Laboratory of Developmental Psychology on the subject of "malnutrition and behavior." In the course of the seminar I sought to clarify the significance of the problem—practical but, especially, theoretical; reviewed some of the important studies; and offered to interact with the relevant personnel of the Laboratory following the seminar, with emphasis on the Indian project.

Eventually, the study was designed and approved by the NIMH review boards. Unfortunately, it proved to be impossible administratively to implement the proposed Indian study. Yet at that time both Dr. Radke-Yarrow and Dr. Barrett were already "hooked" on the idea of a field study on malnutrition and behavior in children and searched for another appropriate study population.

Their concern came to my attention during the summer of 1977. To me, the solution of the problem was immediately clear: the study should be carried out in Latin America and, more specifically, in Guatemala.

I visited Guatemala's INCAP (Institute of Nutrition of Central America and Panama) in 1965 as a Temporary Adviser to the World Health Organization's Division of Nutrition, in connection with a project involving the use of longitudinal growth data as a criterion of changes in the nutritional status of populations. When I returned in 1967, my central interest was in a longitudinal study on the effects of nutritional supplements, provided to undernourished children, on mental development. The study was in its early stages.

The first, exploratory steps were taken in 1964/65. Wisely, in view of the complexity and magnitude of the project, three years were devoted to the refinement of the concept of the study, organization of the staff, acquisition of the financial resources, field and administrative reconnaissance, and the development of standard operating procedures. Initially the project was directed by Cipriano A. Canosa (1968), while the psychological aspects of the study were the responsibility of Dr. Robert E. Klein (1979), who eventually came to play a crucial role in the history of the Barrett-Yarrow study.

In 1976 Dr. Klein and I took part in a working conference held in Cali, Colombia, the goal of which was to obtain dependable, detailed, comparative information on the ongoing and recent longitudinal studies on the effects of malnutrition and nutritional supplementation. The report was published a year later (Brožek, Coursin, & Read 1977) and dealt with the Guatemala study as well as five other studies. The information included the starting and the terminal dates for collecting the data, data analysis, and the preparation of the final report; research design, population, research diets, and behavioral stimuli, if any; methods and measurement; and tentative conclusions. Thus I was well informed not only about the study carried out by Dr. R.E. Klein and his co-workers in Guatemala but also about the investigations carried out in other parts of Latin America (Colombia and Mexico).

When asked by Drs. Radke-Yarrow and Barrett about a possible alternative to India, I did not hesitate for a moment to recommend

Guatemala. The data collection bearing on nutritional supplementation and its effects on mental development was completed, but some of the children who had participated in the longitudinal study were enrolled in the local schools and would be available for study. Importantly, the amount of information at hand was impressive. It included detailed data on the food supplements consumed by the children in their infancy and the preschool years, on physical growth, and on mental development.

In principle, my suggestion was accepted. Without delay I contacted Dr. Klein, explained the situation in which my colleagues found themselves, and stressed their professional competence, the seriousness of their commitment to the project and, most importantly, the innovative methodology they proposed to use. It was to deal with a facet of psychological functioning that had not been used in the existing studies on the effects of chronic malnutrition and nutritional supplementation.

Dr. Klein was interested in the proposed collaborative venture. Dr. Barrett called INCAP in September 1977 and asked if he could visit the Guatemala Institute in order to explore, in detail, the possibilities for carrying out the study. He made the pilgrimage to Guatemala later in the month and, in cooperation with Dr. Klein, laid the groundwork for the study.

Between September 1977 and July 1978 the study proposal was refined, and the observational and testing procedures for the behavioral assessment of the children in small groups were developed and field-tested. Dr. Barrett was ready to begin the fieldwork and left for Guatemala in October 1978. The data collection was complete in summer 1979, and Dr. Barrett immersed himself in the analysis of the voluminous data.

The first report on the study appeared in print three years later (Barrett, Radke-Yarrow, & Klein 1982). It was followed by several communications dealing with the conceptualization of the problem, methods of assessment, and the results (Barrett 1982, 1984a, 1984b; Barrett & Yarrow 1985), supplemented by chapters on methodological issues (Barrett 1984c, 1984d).

But what about the co-author of the book, Dr. Deborah Frank? I made contact with her in connection with an international symposium—a "symposium at a distance"—in 1983 (Brožek & Schürch 1984). Dr. Frank, a pediatrician, was asked to comment

(Frank 1984) on Dr. Barrett's quasi-experimental study (Barrett 1984a). However, the commentator and Dr. Barrett were not total strangers. They met in 1979 when Dr. Frank was both a Clinical Fellow in Pediatrics at the Harvard Medical School and a Fellow in the Child Development Unit (CDU) of the Children's Hospital Medical Center in Boston. At that time Dr. Barrett taught child development and research methods/statistics to the CDU Fellows.

Deeply interested in children who "fail to thrive," Dr. Frank in 1981 became Director of the Failure to Thrive Program as well as Staff Physician of the CDU, Boston City Hospital. She has to her credit several publications on the subject (Frank 1986a, 1986b), including the co-authorship of a paper on anthropometric evidence of malnutrition among low-income children (Guyer et al. 1985/1986). Dr. Frank is the technical advisor for the Physicians Task Force on Hunger and a member of the Mayor's Committee on Hunger, in Boston.

Already as a Fellow at the CDU at Children's Hospital, Dr. Frank became concerned with the effects of malnutrition on brain and behavior. Thus, when in 1985 Dr. Barrett wished to strengthen the introductory chapter of the book as well as the sections on the effects of malnutrition on the brain, failure to thrive, and intrauterine growth retardation, it was only natural that he would seek Dr. Frank's cooperation. This was a happy choice. The fruit of the collaborative effort is a significant contribution to the world literature on malnutrition and behavior.

Josef Brožek
Department of Psychology
Lehigh University

References

Barrett, D.E. (1979) Relation between aggressive and prosocial behavior in children. *Journal of Genetic Psychology* 134:117–118.

——— (1982) An approach to the conceptualization and assessment of social-emotional functioning in studying nutrition-behavior relationships. *American Journal of Clinical Nutrition* 35:1222–1227.

——— (1984a) Malnutrition and child behavior: Conceptualization, assessment, and an empirical study of social-emotional functioning. In J. Brožek and B. Schürch (Eds.), pp. 280–306. Repr. in J. Brožek (1985), pp. 280–306.

——— (1984b) The effects of undernutrition on children's behavior. In D.H. Enstrom (Ed.), *Proceedings: Nutrition and Children's Behavior*. Princeton, N.J.: Office of Education, Department of Human Services, pp. 1–49.

——— (1984c) Methodological issues in nutrition intervention research. In J. Brožek and B. Schürch (Eds.), pp. 585–596.

——— (1984d) Methodological requirements for conceptually valid research studies on the behavioral effects of malnutrition. In J.R. Galler (Ed.), *Nutrition and Behavior*. New York: Plenum Press, pp. 9–36.

——— (1986) Behavior as an outcome in nutrition research. *Nutrition Reviews* (Supplement) 44: 224–236.

——— and Yarrow, M.R. (1977) Prosocial behavior, social inferential ability, and assertiveness in children. *Child Development* 48:475–481.

——— and ——— (1985) The effects of nutritional supplementation on children's responses to novel, frustrating and competitive situations. *American Journal of Clinical Nutrition* 42:102–120.

———, ———, and Klein, R.E. (1982) Chronic malnutrition and child behavior: Effects of early caloric supplementation on children's social and emotional functioning at school age. *Developmental Psychology* 18:541–556. Repr. (1984) in S. Chess and A. Thomas (Eds.), *Annual Progress in Child Psychiatry and Child Development*, 1983. New York: Brunner/Mazel, pp. 403–409.

Brožek, J. (1985) *Malnutrition and Human Behavior: Experimental, Clinical, and Community Studies*. New York: Van Nostrand Reinhold Co.

———, Coursin, D.B., and Read, M.S. (1977) Longitudinal studies on the effects of malnutrition, nutritional supplementation, and behavioral stimulation. *Bulletin of the Pan American Health Organization* 11:237–249.

———, and Schürch, B. (Eds.) (1984) *Malnutrition and Behavior: Critical Assessment of Key Issues*. Lausanne, Switzerland: Nestlé Foundation for the Study of Nutrition Problems in the World, Publication Series Vol. 4.

Canosa, C.A. (1968) Ecological approach to the problems of malnutrition, learning, and behavior. In N.S. Scrimshaw and J.E. Gordon (Eds.), *Malnutrition, Learning, and Behavior*. Cambridge, MA: MIT Press, pp. 389–397.

Frank, D.A. (1984) Malnutrition and child behavior: A view from the bedside. In J. Brožek and B. Schürch (Eds.), pp. 307–326.

——— (1986a) Biological risks in nonorganic failure to thrive: Diagnostic and therapeutic implications. In D. Drotar (Ed.), *New Directions in Failure to Thrive: Implications for Research and Practice* New York: Plenum, pp. 17–27.

———, Alley, D., and Brown, J.L. (1986b) Primary prevention of failure to thrive: Social policy implications. In D. Drotar (Ed.), *New Directions in Failure to Thrive: Implications for Research and Practice*. New York: Plenum, pp. 337–359.

Guyer, B., Anderka, M., Friede, A., Bithoney, W., Frank, D.A., Fogarty, S. and Wehler, C. (1985/1986) Anthropometric evidence of malnutrition among low income children in Massachusetts in 1983. *Massachusetts Journal of Community Health* 2.

Keys, A., Brožek, J., Henschel, A., Mickelsen, O., and Taylor, H.L. (1950) *The Biology of Human Starvation*. Minneapolis: University of Minnesota Press.

Klein, R.E. (1979) Malnutrition and human behavior: A backward glance at an ongoing longitudinal study. In D.A. Levitsky (Ed.), *Malnutrition, Environment, and Behavior: New Perspectives*. Ithaca, NY: Cornell University Press, pp. 219–237.

Yarrow, M.R., and Waxler, C.Z., with the collaboration of Barrett, D.E., Darby, J., King, R., Pickett, M., and Smith, J. (1976) Dimensions and correlates of prosocial behavior in children. *Child Development* 47:118–125.

PREFACE

Why study the effects of malnutrition on behavior? We have heard objections from two sides. On the one hand, we have been told, it is obvious that being underfed is bad for children. If children receive insufficient food, they will grow poorly and will act differently from children who do receive adequate food. Rather than study the possible adverse effects of undernutrition, why not simply find out who is underfed and then feed them? On the other hand, there is the argument that there is no sense pursuing this line of inquiry because after more than two decades of research, there is no definitive study showing that malnutrition is a cause of impaired behavioral development. The reason for this, proponents of this view argue, is that it is impossible to determine the effects of malnutrition *per se* on children's development. If you study free-living communities who are underfed and compare them to well-fed communities, the groups will differ on innumerable characteristics which themselves are predictive of later behavior. And if, within an endemically malnourished population, you compare the most and least malnourished, you will find that the most poorly-nourished will have been treated differently from the better-nourished children in a multitude of ways, both in and outside of the home.

To some degree, critics are correct. We know that malnutrition is undesirable, and even after 20 years of research study, the empirical associations between malnutrition and behavior that have been uncovered raise many interpretive issues. Nutrition and behavior become associated, most likely, by multiple pathways, some of which may be mediated by extranutritional variables.

However, these concerns should encourage rather than dissuade us from research. The fact that we know that a problem is serious should not prevent us from studying its antecedents, correlates, and consequences. In fact, the pervasiveness of undernutrition makes more urgent the task of documenting the conditions under which it occurs and the conditions under which it is likely to have its most serious consequences.

The problem of "confounding" of nutritional variables with extranutritional variables is not a problem specific to nutrition research. Every type of pathology or environmental stress which occurs in the real world occurs under certain conditions; that is, it has environmental correlates. Whether we are studying an organic insult such as premature birth, a social stress such as parental neglect, or a nutritional environmental deprivation such as protein-calorie malnutrition, each type of problem tends to occur under certain circumstances and to not occur under others. Rather than preventing us from studying the problem, this knowledge should encourage us to (a) identify the conditions that tend to co-occur with malnutrition and (b) determine the conditions under which malnutrition may have its most adverse and least adverse effects.

In this book we examine what is known about the effects of malnutrition on children's behavior. We attempt to show that contrary to the belief of many scientists and policymakers, the study of the effects of malnutrition is not beyond our capabilities. Moreover, the results of a great number of research studies using a variety of procedures and populations provide us with consistent data. We believe the following conclusions are supported. First, the effects of malnutrition on behavior are likely to depend on a number of conditions of a person's life—not only the age at which malnutrition occurs and its duration and severity, but also the response of the caregiving environment to the malnourished person. Second, and most important, when malnutrition does adversely affect behavior, the nature of the effects is predictable and describable. Children with a history of undernutrition show long-term changes in behavior, including changes in activity, affect, social responsiveness, and involvement with others. These behavioral effects are among our most sensitive indicators of long-term food deficit.

We have organized this book into five chapters. In Chapter One we discuss the question of how we define undernutrition when we conduct research on malnutrition and behavior. There are different ways of conceptualizing and operationalizing undernutrition. In this chapter, we discuss the major approaches, their assumptions, objectives, strengths, and limitations. We also examine the estimated prevalence of malnutrition in the United States and abroad.

In Chapter Two we discuss the methodological issues in conducting research on malnutrition and behavior. First, we discuss the relationship between research design, operationalization of the independent variable nutritional status (or nutritional history) and the theoretical questions that will be addressed by the research. It is often not recognized that selection of the independent variable places constraints on the research design and thus on the theoretical inferences that can be drawn from the study. Conversely, selection of a design makes possible certain operationalizations of the independent variable(s) and precludes others. Both factors affect the research issues that can be addressed. Second, we discuss different approaches to measurement of behavior in studies on malnutrition and behavior. We have seriously underestimated our abilities to measure important behavioral outcomes. We discuss an approach to the conceptualization of social and emotional functioning and also the specifics of measurement operations. Third, we discuss the threats to the validity of research studies on malnutrition and behavior, giving particular emphasis to threats to the 'construct validity' of our interpretations—the issue of whether our inferences are consonant with the empirical relations identified and, in particular, the operations we have used to obtain these relations. Finally, we review the different classes of research design and their characteristics, including a discussion of the situations in which they may or may not be useful.

In Chapter Three we review the research on the effects of malnutrition on behavior. Our purpose is to convey the thrust of the findings that emerge from different sources; that is, from different types of designs and study populations. After discussing the early clinical observations of malnourished infants which provided a basis for the experimental research, we discuss first the experimental research on malnutrition in animals and second, the research on the effects of malnutrition on human behavior and development. The results of the animal research are presented in detail because they suggest the dimensions of behavior that may be sensitive to severe undernutrition and indicate the role of the environment in moderating the effects of undernutrition. The human research is reviewed in categories, each corresponding to a different type of population and study design. The design characteristics of each type of study are also discussed here, in the introduction to each review of literature, and they are related to the general categories of research design identified in Chapter Two. We emphasize that this review is a focused one. The purpose is to show that different sources of data converge to suggest specific behavioral functions that are affected by both severe and subclinical malnutrition, so that we can make predictions about what behaviors will be most vulnerable to malnutrition in the developing child.

Chapter Four is a presentation of research recently completed by the first author on the effects of chronic malnutrition on children's behavior at school age. This study is the first to systematically examine the effects of early and chronic undernutrition on children's capacities to solve problems, interact with peers, and show normal social and emotional responsiveness to others later in childhood. We give particular attention to the construct validation of the measures developed for studying the social and emotional development of children. The concluding section of Chapter Four elaborates a framework for thinking about the long-term and cumulative effects of food energy deficit in children, and provides support from recent clinical intervention studies.

Chapter Five includes a discussion of the present research, an attempt to integrate the present findings with previous work, and a discussion of as-yet unresolved issues in the research on malnutrition and behavior. We consider the reliability of the present findings, their relation to other findings on social and emotional consequences of undernutrition, and the viability of an interactive model for understanding these effects. We also consider the question of sensitive periods for nutritional deprivation or supplementation, and the relative vulnerabilities of different aspects of behavior function to undernutrition. The social implications of the research are discussed, and areas of current research on malnutrition and behavior are highlighted.

Many people helped make possible the research I conducted in Guatemala on the behavioral effects of malnutrition, which is reported in Chapter Four. First, I would like to thank Dr. Marian Radke-Yarrow and Dr. Josef Brožek.

Dr. Radke-Yarrow was my supervisor at the National Institute of Mental Health when I carried out the study and was co-investigator. In her distinguished career, Dr. Radke-Yarrow has made major contributions to methodology for studying children's social behavior and affect, and much of the conceptual groundwork for this study came from her work. I am indebted to her both for her professional supervision and her personal support.

Dr. Brožek is one of the most eminent researchers and theoreticians in the area of malnutrition. He was a consultant to this study at its inception, and he encouraged us to pursue vigorously our interest in focusing on social and emotional sequelae of undernutrition. It was Dr. Brožek who made possible our involvement with the Institute of Nutrition of Central America and Panama (INCAP) through his correspondence with Dr. Robert E. Klein. Throughout the study, including the data analysis and the reporting of our results, Dr. Brožek has been a thoughtful critic of our research, and he has been a valued friend and colleague.

I would also like to thank Dr. Robert E. Klein, co-investigator in the Guatemala research. Dr. Klein, Chief of the Division of Human Development at INCAP, provided the background data for the Guatemala subjects, including nutritional supplementation data, for the present analyses. This study would not have been possible without his cooperation. In addition, Dr. Klein provided invaluable advice and assistance while I was in Guatemala conducting the data collection.

A number of people were involved in the data collection and analysis. I would like to thank Miriam Mendietta, Gloria E. Valdez, and Rosa Eugenia Hernandez, who collected the data on child behavior in Guatemala. Claire Horowitz and Dr. Brian McLaughlin from NIMH and Dr. Joel Hoffman of Children's Hospital in Boston conducted the computer analyses. I am grateful for their help. Finally, I would like to thank Marie Donovan, who typed the manuscript and helped with text editing.

David E. Barrett

FIGURES

TABLES

Chapter One: The Problem of Undernutrition: Definition, Assessment, and Prevalence

1.0 INTRODUCTION AND OVERVIEW OF THE CHAPTER

The need for food is among the most basic of human needs. Adequate food is needed in order to survive, to grow and reproduce, to carry out physical work, and to learn from one's experiences. A large percentage of the world's population, and in particular the world's children, nevertheless receives inadequate calories and protein to maintain healthy growth and development.

The pervasiveness of undernutrition in today's world is staggering. In many parts of the world, more than half the child population is inadequately nourished and at risk for severe malnutrition. Even in the United States, where there is greater concern with obesity and specific nutrient deficiencies, a substantial proportion of our population is underfed. The problem in this country is particularly common among pregnant and lactating women. For women living at or below the poverty level, the chances of giving birth to a prenatally undernourished child are disturbingly high.

Yet despite the awareness of scientists, citizens, and governments of the pervasiveness and seriousness of worldwide undernutrition, surprisingly little has been written and disseminated concerning the consequences of undernutrition for the developing child. The bulk of the research has focused on nutrition and growth. More recently, studies have attempted to relate nutritional history to intellectual development. What has been lacking is a systematic investigation of the effects of undernutrition on children's behavior: the child's capacity to seek and learn from experiences, and to be emotionally capable of leading a happy and productive life.

Before addressing the question of how malnutrition may influence behavior and development, it first is necessary to consider how it is that we determine whether a person is undernourished.

Malnutrition is a complex abstract concept, which may be operationally defined in various ways. The definitions chosen inevitably influence both research design and the theoretical inferences that can or cannot be made about the relationship between malnutrition and behavior. Each mode of operationalizing the concept of malnutrition has merits and limitations in approximating the ideal abstraction that must be understood before findings can be interpreted. In this chapter we examine the ways that investigators in this field have defined current nutritional status and inferred previous nutritional history in their subjects.

For the purposes of this discussion, "malnutrition" will be equated with undernutrition, deficits in the quantity and/or quality of nutrients available to the organism, without considering problems of nutrient excess, which some authorities have also labelled "malnutrition". Moreover, we will focus primarily on defining protein-energy malnutrition. Both early, severe protein-energy malnutrition and chronic, subclinical protein-energy malnutrition will be considered. We will not address the issue of whether one can define malnutrition by measuring deficiencies of micronutrients, such as iron, which may, in conjunction with protein-energy malnutrition, or alone, produce adverse behavioral effects. For a discussion of the identification of micronutrient deficiencies and their effects on behavior, topics outside the scope of this book, the reader is referred to current reviews (Balasz, Lewis, & Patel, 1979; Lozoff et al., 1985).

Although recent theory suggests that behavior itself is a highly sensitive indicator of nutritional status (Solomons, 1985), this chapter will consider measurement and interpretation only of the nonbehavioral parameters by which malnutrition may be identified. We will focus on those indicators that have been widely used in the identification of child populations at nutritional risk and in studies of malnutrition and behavior, excluding from discussion measures of immunocompetence that have been used to assess nutritional risk of infectious morbidity (Solomons, 1985), and other measures that are primarily relevant to the study of nutritional status in adults, such as work productivity, sperm count, or dark adaptation (Solomons, 1985).

The indicators of malnutrition to be reviewed in this chapter fall into two major categories: those related to nutritional intake determined by history or by observation, and those related to physical growth, both pre- and postnatal. For each specific indicator in both categories there are general methodologic issues that must be considered. These include:

1. To what extent a given indicator is specifically a manifestation of nutritional status and, conversely, what non-nutritional processes may be contributing to a given indicator;

2. Within what limits the indicator is susceptible to accurate measurement in a research setting;

3. To what reference standards the indicator should be compared;

4. What statistical cutoffs of the reference standards should be chosen to classify an indicator as representative of adequate or inadequate nutrition;

5. What methodologic merits and limitations are unique to a particular indicator.

In this chapter, these methodologic issues will be discussed for each indicator of nutritional status used in the investigation of the relationship of malnutrition and behavior. In a concluding section, we briefly summarize the prevalence of childhood malnutrition in the developing world and in the United Status according to the indicators described in the chapter.

2.0 INDICATORS OF MALNUTRITION

2.1 *Dietary Intake as an Indicator of Nutritional Status*

2.1.1 Specificity

Dietary intake, by definition, is an exclusively nutritional parameter, whereas other indicators used to classify nutritional status, such as growth, attained size, or circulating nutrients, express the composite effects of both nutritional and non-nutritional factors. The purest measure of nutritional status would be a summary of the total amount of nutrients ingested by the mother during pregnancy and lactation and by the child from birth to the time of behavioral assessment (Klein, 1979). One could then correlate variations in the quality and quantity of the mother's or child's diet in the natural setting with the child's behavior without a priori assumptions that a certain set of dietary practices will or will not have adverse effects. Unfortunately, no such accurate summary measurement of dietary intake is obtainable with current techniques.

In free-living populations, the methods of documenting dietary intake over time that are feasible can only approximate this ideal measurement of nutritional status.

There are three possible modes for evaluating dietary intake: direct observation, dietary histories, and examining individual differences in intake of nutritional supplements. The first two techniques entail serious measurement difficulties, as will be described in the next section, but are not greatly confounded by possible non-nutritional factors that may influence observed behavioral outcomes. In contrast, while supplementation designs permit precise measurement of supplement intake, differences among subjects in the amount of supplement ingested cannot be considered solely a nutritional indicator, for two reasons. First, supplementation studies involve frequent interpersonal contact between research personnel and the study sample, thus introducing extra-nutritional social stimulation of parents and children that might influence outcome. Secondly, there may be unmeasured motivational differences between subjects who choose to obtain supplement regularly for themselves or their children and those who do not, so that differences in observed outcomes may reflect differences in parental characteristics, rather than the effects of the nutritional supplement alone. These issues are discussed in greater detail in Chapters Two and Three.

2.1.2 Accuracy of measurement

Precise techniques for measurement of dietary intake exist. All involve weighing foods before they are eaten and then performing laboratory analyses of duplicate samples (Acheson et al, 1980). Unfortunately, such meticulous observational techniques are not applicable to the field conditions under which most research on undernutrition and child behavior is conducted. There are two major reasons. First, most subjects who participate in typical field studies will not be so highly motivated as to comply with the rigorous scientific requirements of an observational study, which involves great time and effort. Second, intense scrutiny of food practices is likely to alter usual dietary behavior. Thus, findings from such a study would not generalize to the home situation.

For these reasons, investigators in field studies often use historical rather than observational methods to determine the home diets of pregnant women and children. Because studies of undernutrition and child behavior are usually conducted in semiliterate populations, interview rather than dietary methods are used.

The majority of investigators in field studies of undernutrition and child behavior have used the 24-hour dietary recall, where the informants are asked a structured series of questions designed to identify the identity and quantity of all foodstuffs and beverages ingested by themselves or their child in the 24 hours prior to the interview. The 24-hour dietary recall is relatively quick to obtain and is considered less subject to memory distortion than the three- or seven-day dietary recall. However, any dietary history method for determining subjects' nutritional status is subject to a number of difficulties. To simplify the discussion we will use the 24-hour dietary recall as illustrative since it is so widely used. However, the problems intrinsic to the 24-hour recall are also found for the other dietary history methods, including three- and seven-day recalls, and food frequency studies, particularly when applied to women of child-bearing age and children.

The first issue that arises is whether the dietary recall is valid, that is, whether subjects accurately report their intake or their child's intake. The issue has been studied only in adult populations. When 24-hour recall data are compared to intake observed in the same 24-hour period, investigators have consistently noted that there is a tendency for respondents to overestimate small intakes and underestimate large ones (Acheson et al., 1980; Gersovitz, Madden, & Smicklas-Wright, 1978; Krantzler et al., 1982), as well as a tendency to alter responses in the direction of social desirability (Madden, Goodman, & Guthrie, 1976). Discrepancies between reported and observed intakes of energy may be as high as 21% (Acheson et al., 1980), even when the subjects are themselves members of a scientific team. The accuracy of dietary recalls when compared to the observed intakes of young children has never, to our knowledge, been formally assessed. The accuracy of such recall depends not only on the respondents' memory and motivation, but on their ability to convey estimates of quantity, an ability that may be limited in young children, and also in poorly educated adult caretakers.

Even if the accuracy of a 24-hour dietary recall is considered adequate, any one day's intake may be poorly representative of a subject's cumulative nutritional experience over time. Typically, dietary intakes of adults have, in industrialized countries, been found to be higher on weekends than on weekdays (Gersovitz et al., 1978), and in developing countries to be lower before harvest seasons (Prentice et al., 1980). In addition to these cyclical temporal differences in 24-hour dietary intakes, for infants and children there also are normal developmental changes over the early years of life as the child progresses from breastmilk or breastmilk substitutes, to pureed foods, and then to the family diet. The intake of a child at any given age cannot be automatically extrapolated to intake even a few months later. Similarly, because of biologic changes in appetite and nutritional requirements over the trimesters of pregnancy, any single dietary recall may be poorly representative of a mother's nutritional experience throughout pregnancy (Picone et al., 1982).

To obtain a more representative picture of cumulative dietary intake over time, investigators of malnutrition during pregnancy and childhood have obtained repeated 24-hour dietary recalls. When multiple recalls are obtained, test-retest reliability coefficients are found to be higher for dietary components habitually consumed from staple foods, such as energy, than for micronutrients (Rush & Kristal, 1982). Repeated recalls are considered essential for classifying the nutritional status of individuals (Hubbard & Hubbard, 1985), but even so the intercorrelations between multiple recalls for a single individual are not robust.

For example, in the 1969–1977 longitudinal study conducted by the Institute of Nutrition of Central America and Panama, INCAP, in Guatemala (see Klein, 1979), the median intercorrelations between two recall estimates for dietary energy taken months apart ranged from .12 for children 18–60 months, to .47 for mothers during pregnancy. In recent studies in the United States, two to six repeated dietary recalls during pregnancy were used, resulting in reliability coefficients ranging from .43 to .67 (Rush & Krystal, 1982), and intrasubject coefficients of variation (intrasubject standard deviation of caloric intake/mean caloric intake X 100) ranging from 19.9 to 21.4 (Picone et al., 1982). This low- to moderate degree of consistency in dietary recall data reflects both the inaccuracies inherent in dietary histories and true differences in intake over time.

For characterizing differences between individuals, multiple recalls are superior to single recalls. We emphasize that multiple recalls cannot fully overcome the uncertainties inherent in dietary history methods for identifying individuals as well or poorly nourished (Hubbard & Hubbard, 1985). However, dietary recall methods are considered appropriate for classifying the relative nutritional status of populations, since in large samples in which only mean values of intake for a given group (e.g., black children under 5) are considered, the numerous possible errors in estimating dietary intake are assumed to cancel each other out (Hubbard & Hubbard, 1985).

Once dietary histories have been determined, whether for individuals or populations, another methodologic difficulty arises: How to determine accurately the nutrient composition of the reported intake. As mentioned above, laboratory analysis of duplicate samples of food eaten is the most precise method for determining the nutrient composition of a given meal, but is impractical for field settings (Acheson et al., 1980). In most cases the nutrient composition of recalled dietary intakes is determined from standardized food tables, such as those established by the United States Department of Agriculture (Church & Church, 1970). However, food tables are based on the average composition of a number of different samples of a given food item (such as a slice of bread), and may not accurately reflect the composition of the particular example of that food item ingested by the subject. Furthermore, it is particularly difficult to ascertain from food tables the nutrient composition of composite dishes such as stews or casseroles, where the exact amount of each constitutent of the composite dish is hard to estimate (Acheson et al., 1980).

Thus there are three areas of difficulty in interpreting dietary recall data: inaccurate reporting, lack of stability of intake over time, and uncertainty inherent in determining nutrient content of reported intake from standardized food tables.

In order to overcome the uncertainties intrinsic to classifying nutritional status by the quality and quantity of home diets determined by dietary history methods, a number of investigators have chosen to identify the effects of malnutrition on behavior by assigning subjects thought to be suffering from nutritional deficits to various forms of nutritional supplementation during pregnancy, lactation, and/or infancy and early childhood. The logic of the nutritional supplementation approach is as follows. If persons from a nutritionally at-risk population are comparable in their home diets with respect to the nutrient(s) that is limiting in their diets (e.g., protein), then, by supplementing individuals (or groups) differentially on that nutrient, individual differences in nutritional status can now be tied to degree of supplementation.

Nutritional supplementation designs avoid some of the difficulties described above in measuring home diets—in most cases, the composition of the supplement is precisely known for all nutrients, and changes only if the investigator chooses to alter its composition. Moreover, if subjects ingest the supplement under the observation of the research team, as occurred in the INCAP study (Klein, 1979), the Bacon Chow study (Adair & Pollitt, 1985), and the Gambian study (Prentice et al., 1980), precise quantification of amounts received is possible. In other studies, where the supplement is distributed for home use (Evans et al., 1980; Mora et al., 1979), or provided as school meals (McKay et al., 1978), the investigators can define the quality and quantity of the supplement provided but cannot be certain about the amount ingested by the target subject (as opposed to being shared with other people, sold, or wasted).

Thus, in spite of the accuracy with which the nutrient quantity and quality of supplements can be determined, supplementation studies do not permit accurate classification of nutritional status in a way that can unequivocably be generalized to unsupplemented populations. Further, in supplementation designs, the investigator cannot evaluate how much of a given nutrient a subject receives in total, but only the amount received in the supplement. As discussed above, it is impossible to quantify accurately the amount of home diet ingested in addition to the supplement. In addition, one cannot assume that the total amount of supplement supplied serves as net increment to the subjects' diet, even if supplement intake is observed by the investigators. To a certain degree, supplement intake acts as a substitute rather than an increment to usual diet. For example, if a child has been fed at a supplementation center, the family may divert the food usually fed to that child at home to other family members, or use the economic resources usually applied to food purchase for other goods or services. Beaton (1982) has estimated that, in general, whether supplement is supplied at feeding centers or at home, approximately 50% represents a net increment to the target child and 50% is either substituted for previous diet or diverted to other family members. A final concern, and one we address further in Chapters Two and Three, is that one cannot assume that the positive nutritional consequences of supplying remedial amounts of a nutrient will mirror precisely the adverse effects of an analogous degree of deficit of that nutrient. Only a research design based on systematic nutrient deprivation of initially replete subjects could accurately address this issue. Although such studies have occasionally been conducted in consenting adult males (Keys et al., 1950), they would be ethically unthinkable in pregnant women or in children.

Thus, although the quality and quantity of nutrients ingested from supplement can be more accurately determined than the quality and quantity of nutrients in home diets, as measured by dietary histories, supplementation studies do not permit classification of subjects as well or poorly nourished, but only as well or poorly supplemented.

2.1.3 Reference standards

Once home diet or supplement intake has been ascertained within the limits of measurement accuracy, reference norms are used to classify intake as adequate or inadequate. National standards for dietary intake vary from country to country (Wretlind, 1982). Although some investigators of malnutrition and child behavior have used local standards (Herrera et al., 1980), the majority have evaluated dietary intake according either to FAO/WHO guidelines published in 1973 (FAO/WHO, 1973) or to Recommended Dietary Allowances (RDAs) devised by the National Research Council of the National Academy of Sciences in the United States (NRC, 1980). These standards are not static, but undergo periodic revisions as new information becomes available. Therefore, a pattern of dietary intake that was judged adequate or inadequate by standards accepted at the time a given research study was conducted might be reevaluated if reanalyzed according to current norms.

The FAO/WHO and NRC RDAs also differ systematically from each other, in areas relevant to the study of undernutrition and child behavior. For example, the 1980 NRC recommendations for energy intake for infants are identical to the 1973 FAO/WHO recommendations, but those for children are lower (NRC, 1980). Conversely, NRC recommendations for protein intake by pregnant women far exceed the FAO/WHO standards (NRC, 1980). The differences between the two standards are not, then, in any consistent direction, but differ by the nutrient and the category of persons being considered. Depending on the norms selected by the investigator, dietary intake data would classify some individuals as adequately nourished who at the same intake might be classified as poorly nourished by other norms. As of this writing, there are no accepted conventions as to which set of norms should be used in studies of undernutrition and child behavior (or any other nutritional issue, for that matter). Obvious difficulties arise in attempting to generalize from studies of dietary intake evaluated according to one set of norms to those using another, since there inevitably will be a gray zone where members of one sample, classified as poorly nourished, would be considered well-nourished in a different sample using different norms.

In addition to the interpretive difficulties arising from discrepancies between the FAO/WHO and NRC norms at any given epoch and the discrepancies within each set of norms over time, there are methodologic uncertainties common to both. To understand these problems, we must review briefly the conventions underlying the evolution of these norms.

First, although standards for nutrient intake are expressed as "Recommended Daily Allowances" (RDAs), they are designed to evaluate not the adequacy of a single day's intake but, rather, of average daily intakes. The concept of daily allowance is simply a convention of presentation "in order to simplify dietary calculations" (NRC, 1980), not a standard applicable to intake in any one 24-hour period. In fact, the NRC recommends that the RDAs should be used to evaluate average intakes over a five- to eight-day period (NRC, 1980). An important implication is that if subjects are classified as well or poorly nourished on the basis of a single day's intake compared to the RDAs, misclassifications of nutritional status may occur because of failure to consider compensatory or insufficient intake on other days.

Next, the major interpretive difficulty in using FAO/WHO or NRC norms for assessing the adequacy of dietary intake arises because of the uncertainty inherent in extrapolating information regarding nutrient needs derived primarily from studies of populations in industrialized countries to populations of different ethnic backgrounds living under different environmental stresses.

In general, recommended intakes are specified for age, sex, body size, and whether or not the subject is pregnant or lactating (NRC, 1980). However, the recommended intakes are designed for the maintenance of already adequate nutritional status, which cannot be assumed in the deprived populations in which undernutrition and child behavior are usually studied (Beaton, 1985). Further, the recommended intakes are based on the assumption that they must support only light physical activity.

Nutritional needs of persons living in developing countries or in impoverished areas of the industrialized world may be very different from those of persons in more developed areas. In the less well-off populations, cumulative nutritional deprivation of young children may evolve because of the interaction between underfeeding and recurrent infections. Malnutrition impairs immune function (Suskind, 1981). With each episode of infection, the child's appetite and intake decrease, while losses of nutrients increase from the metabolic cost of response to infection as well as from diarrhea and vomiting (Scrimshaw, 1981). Thus, unless the child receives compensatory nutritional intake following infection, each infection leaves the child more malnourished and more vulnerable to future severe and prolonged infections, which in turn will further impair the child's nutritional condition. The standards by which dietary intakes are evaluated are based on the observed intakes of healthy children in the developed world and do not adequately reflect the needs of impoverished children burdened by recurring infectious morbidity. These children must not only maintain their nutritional status but repair deficits. Thus, classifying the nutritional status of poorer children by using standard RDAs to evaluate their intake probably underestimates the severity and prevalence of their nutritional deprivation.

Although most authorities agree that RDAs established for populations in industrialized countries underestimate the needs of populations in the developing world (Beaton, 1985), some investigators have argued that humans may be able to adapt to intakes chronically well below the average requirement by increasing their metabolic efficiency without either depleting their body stores or diminishing their activity (Sukhatme & Margen, 1982). This assertion has not been validated empirically in any behavioral research, although there are data from Gambia that breastfed infants maintain their growth trajectories in the first three months of life in spite of levels of maternal and infant dietary intake that are well below the RDAs (Prentice et al., 1980). Uncertainties regarding the adaptive capacity of any given subject to maintain energy homeostasis at intakes below those described in standard norms make it impossible to differentiate on the basis of dietary intake those whose undernutrition has impaired function from those who have adapted to the lower intake and are not impaired.

In summary, two sets of widely used norms exist for evaluating the adequacy of dietary intake—those published by FAO/WHO (1973) and those published by the National Research Council of the National Academy of Sciences (1980). Using these norms permits comparisons between different samples in which the same sets of norms have been used. However, the norms are not precisely consistent with each other, nor with earlier versions of the same norms. Moreover, in using such norms to evaluate the adequacy of dietary intake, one must realize that in the case of children from impoverished populations, one may overestimate the degree of nutritional deprivation by failure to recognize individual adaptive capacity to below-average intake or, more likely, underestimate the severity of nutritional deprivation, by failure to recognize the effects of chronic infectious morbidity and other environmental stresses on nutritional needs.

2.1.4 Cutoffs for evaluating dietary intakes

In selecting criteria for evaluating the adequacy of either home dietary intake or supplementation according to the RDAs it is critical to understand the statistical conventions underlying the published dietary norms (FAO/WHO, 1983; NRC, 1980). For energy and other nutrients, it is assumed that there is biologic variability of requirements among individuals of the same age, gender, body size, and child-bearing status. While the magnitude of normal biologic variability of nutrient requirements cannot be known with certainty, it is conventionally assumed that "true" individual requirements follow a normal distribution with a standard deviation of approximately 15% of the mean requirement (Beaton, 1985).

RDAs for energy are set at the average requirement for a given class of individuals, logically implying that half of the individuals of that class will have a "true" requirement higher than the published recommendation, and half lower. There is a general consensus that at all ages an intake more than two standard deviations below the RDA for energy (i.e., less than 70% of the RDA) indicates clinically significant energy deprivation , since only 2.5% of a population would be expected on statistical grounds to have "true" energy requirements this low (Beaton, 1985; Livingston et al., 1975).

In contrast to energy allowances, the RDAs for protein and other developmentally important nutrients, such as iron, are set at two standard deviations above either the average requirement found in population studies, or some minimum shown in laboratory studies to prevent or reverse deficiency, maintain metabolic equilibrium, or preserve tissue saturations (Wretlind, 1982). Thus, while 50% of healthy individuals will have "true" energy requirements greater than the RDA for energy, only 2 to 3% will have a "true" requirement for nutrients other than energy that is greater than the RDA. Thus different statistical cut-offs than those used for energy are necessary for classifying individuals as nutritionally deficient on the basis of intake of nutrients other than energy. Two approaches to determining cut-offs have been suggested. Livingston et al. (1975) suggest that dietary intakes of more than four standard deviations below the RDA for protein (40% of the RDA) place the individual at nutritional jeopardy, since very few persons will have true requirements this low.

The other approach has been developed for evaluating the nutritional well-being of populations, but not individuals. The method is outlined in detail by Beaton (1985), and uses a probabilistic model based on dividing documented intake of nutrients other than energy into intervals, estimating on the basis of the normal distribution the probability that an intake lying within those intervals would be inadequate, multiplying that probability by the number of subjects observed with that intake to obtain the number of subjects whose intake is "truly" inadequate, summing this number and dividing it by the total number of subjects to determine the proportion of the population whose intake for a given nutrient is inadequate. Again, this approach cannot be used to classify individuals, but only to specify the proportion of a sample whose intake is inadequate.

In summary, it is widely accepted that energy intakes two standard deviations below the RDAs identify individuals who have a high probability of being clinically malnourished. The criteria for classifying nutritional status on the basis of intakes of nutrients other than energy are less clear, and are not always applicable to individual cases.

2.1.5 Merits and limitations

Identification of malnutrition by documenting dietary intake has several advantages. Most methods are relatively rapid, inexpensive, and noninvasive. Moreover, in studies of the effects of malnutrition on children's behavior, classification of subjects on the basis of dietary intake has an important heuristic advantage: it requires no assumption that changes in body size or other health parameters mediate the relationship between nutrition and behavior. If children can be classified on the basis of nutritional intake alone, deficits in activity and learning that may evolve from undernutrition before other structural or functional deficits are noted could be identified (Beaton, 1983; Solomons, 1985).

In spite of the attractive theoretical possibilities that exist if we use "pure" nutritional information to identify malnutrition, there are methodological, including measurement, problems that limit the usefulness of classifications of individual nutritional status on the basis of dietary intake. Direct observation of dietary intake, while accurate, is prohibitively time consuming and probably poorly representative of intake over time. Dietary histories, although easier to perform than observations, are not highly accurate. To characterize the nutritional status of individuals by dietary histories, repeated interviews are required, both because single intakes are poorly representative of intake over time and because the norms by which intake are evaluated are designed to assess the average of multiple daily intakes, rather than a single day's experience.

Even when multiple dietary recalls are obtained, intraindividual variability may exceed interindividual differences, making identification of malnutrition in any one subject questionable.

In general, dietary history methods are most appropriately used to characterize not individuals, but populations. Such a use of dietary recall data best fits the statistical assumptions by which intakes are evaluated and obviates some of the biases introduced by inaccuracies in individual reports of dietary intake and true instabilities in individual dietary behavior over time.

Investigators of malnutrition and child behavior have therefore never used observations or reports of usual diet as the sole indicator of nutritional status. Rather, dietary intake measures are used to identify populations at nutritional risk within which individuals are further classified on the basis of supplementation experience (e.g., Rush, Stein, & Susser, 1980) or anthropometric (e.g., Graves, 1976) criteria.

Classification on the basis of supplementation experience offers improved accuracy over dietary history methods in documenting the quality and quantity of cumulative nutrient intake. Although supplementation studies may introduce non-nutritional factors that may complicate interpretation of results, they make an important contribution to our understanding of the effects of individual differences in nutrient intake on later behavior and development.

2.2 Measures of Physical Status Including Perinatal Parameters

The most widely used indicators of nutritional status in investigations of the effects of malnutrition on children's behavior are those based on deficits of physical size.

The earliest studies of malnutrition and children's behavior (DeLicardie & Cravioto, 1974; Hansen, 1971; Hertzig et al., 1972; Stoch & Smythe, 1963) identified subjects on the basis of clinically obvious deficits in body size, usually leading to hospitalization. Precise measurements were not obtained. The malnourished subjects in these early investigations were suffering from the classic syndromes of nutritional deprivation—marasmus, ascribed to starvation, and manifested by severe emaciation, or kwashiorkor, ascribed to relative protein deficiency, and manifested by depressed body size, skin breakdown, liver enlargement, and edema (Viteri, 1981). Children with these clinical syndromes represent the life-threatening extremes of the spectrum of derangements in body size and composition that may be produced by malnutrition. Marasmus and kwashiorkor indicate physiological decompensation, often precipitated by infection, occurring in children whose nutritional stores have been chronically depleted (Viteri, 1981).

Findings on the behavioral sequelae of clinically recognized marasmus and kwashiorkor cannot a priori be generalized to children who are malnourished but never come to medical attention in a premorbid state.

A more subtle and continuous classification of nutritional status that does not rely on the presence or absence of clinically obvious marasmus or kwashiorkor has evolved, based on anthropometry, the science of body measurements made for the purpose of identifying and categorizing human differences (Johnston & Lampl, 1984). Recent research on malnutrition and child behavior has relied primarily on height, weight, and head circumference as the salient anthropometric indicators of nutritional status (e.g., Galler et al., 1983a; Grantham-McGregor et al., 1983; Graves, 1984). More refined measures of body composition, such as fatfolds and calculations of muscle mass, are used extensively in epidemiological research on nutritional status. Unlike height, weight, and head circumference, however, these body composition measures are not usually evaluated as major predictors of behavioral outcome, and so will not be further discussed.

2.2.1 Specificity

Head circumference, weight, and height differ in the extent to which they may be considered indicators of nutritional status, past and current. Head circumference, which reflects brain growth (Winick, 1976), is a relatively insensitive measure of general nutritional status. Gross alterations in brain size are reflected in head circumference only if the nutritional deprivation occurred in the first two years of life, when the majority of brain growth occurs, and then only if the deprivation was severe and prolonged (Johnston & Lampl, 1984). Moreover, any non-nutritional insult that may impair brain growth, such as asphyxia, infection, or chromosomal abnormalities, will be reflected in head circumference. In sum, depressed head circumference may be indicative of severe nutritional deprivation occurring during the first two years of life. However, without specific exclusion of other possible early insults affecting brain growth, depressed head circumference cannot be assumed to be specific to nutritional deprivation.

Weight, the most nutritionally labile parameter, represents a composite measure of past and present nutritional experience.

Weight should show a steady incremental growth from conception until adulthood.

With nutritional deprivation, weight first fails to increase and then declines. Failure to show expected increments in weight is a subtle and early indicator of current nutritional deprivation, which has been used in Scandinavia and the United States to identify levels of nutritional risk that would go unnoticed in third-world settings (Karlberg et al., 1980; Vietze et al., 1980). However, in most studies of malnutrition and child behavior, diagnosis of malnutrition has depended on the diagnosis of deficits in absolute weight, on the basis of comparison to expected weight for age and sex. A deficit in weight for age alone can identify nutritionally deprived children, but provides no information as to whether the deprivation is concurrent or past.

To assess the chronicity of malnutrition, both weight and height are measured. Unlike weight, height does not decrease with nutritional deprivation, but only fails to increase. Whereas several days of nutritional deprivation will cause weight to fall, height only fails to increase after weeks to months of nutritional deprivation. Thus, a deficit in height may reflect an extended period of nutritional deprivation. By convention, a child with depressed height for age is considered chronically malnourished or "stunted", while a child with depressed weight for height is considered acutely and currently malnourished, or "wasted" (Waterlow, 1972). Many children enrolled in studies of malnutrition and child behavior are both wasted and stunted, showing evidence of acute deprivation imposed upon chronic.

Deficits in body parameters relative to established norms and to each other are used to quantify severity and chronicity of malnutrition. However, it is important to recognize that our ability to identify the timing of nutritional insult from cross-sectional anthropometric measures is limited. As mentioned above, depressed head circumference suggests that the onset of nutritional deprivation occurred in the first two years of life, assuming other causes of depressed head circumference can be excluded. Given a stunted child with depressed height for age, but normal weight for height, it is impossible to determine whether the child suffered an acute episode of nutritional deprivation and then recovered, or whether his height was depressed by chronic low-grade nutritional deprivation without any dramatic episode of weight loss and illness (Zerfas, Shorr & Neumann, 1977). Clearly, the behavioral sequelae of these differing nutritional histories with identical anthropometric presentation may be quite different.

As outlined above, the size and rate of growth of various body dimensions are highly sensitive to nutritional history. Unfortunately, although sensitive to nutrition, deviations in physical growth from expected patterns do not exclusively indicate nutritional experience—they also reflect a variety of pre- and postnatal factors. In studying the effects of malnutrition as manifested by deficits in body size on child behavior, it is important to be aware of the non-nutritional determinants of growth and, when possible, to exclude them as the cause of growth failure before classifying a child as malnourished. These non-nutritional determinants of growth include genetic and perinatal factors, as well as major medical illnesses.

While weight for height, as described above, is generally accepted as a measure of recent nutritional experience, height for age (and thus weight for age) is determined not only by cumulative nutritional experience over time, but by preset, genetic constraints. Given optimal nutrition, different children will attain different heights because of differences in genetic potential, which is conventionally defined by the mean parental height, or the sum of the mother's and the father's height divided by two. Nomograms exist for ascertaining whether a given child's height falls within the range predicted by mean parental height (Hiimes et al., 1985; Tanner, Goldstein, & Whitehouse, 1970). However, such nomograms are useful in determining the child's genetic potential for linear growth only if it can be assumed that variations in the parents' growth express only genetic differences. In impoverished populations, where the parents were probably malnourished in childhood, such an assumption is unwarranted. It is impossible to determine in economically deprived populations to what degree the mean parent height indicates the child's genetic growth potential, as opposed to the parents' own nutritional history. Siblings' height in these settings also may be poorly indicative of genetic potential since the siblings may also have sustained varying degrees of nutritional deprivation. Therefore, in studies of malnutrition and child behavior conducted in economically deprived populations, the genetic as opposed to the nutritional sources of variance in attained height are difficult to quantify, and remain a source of error in classifying children as well or poorly nourished on the basis of height for age.

In addition to genetic determinants of childhood growth, there are congenital and prenatal factors that may influence later growth both in concert with and independently of nutritional experience. The presence of congenital and prenatal constraints on growth can often be inferred from aberrations in the infant's size at birth, although certain non-nutritional perinatal catastrophes such as profound asphyxia, which may adversely influence later growth, can occur in infants of normal size at birth.

The relationship of size at birth to the specificity of growth parameters as nutritional indicators is a complex one. There are two major interpretive issues. The first is to what degree size at birth is itself an indicator of the infant's nutritional experience in utero. The second is whether aberrant size at birth alters the certainty with which patterns of postnatal growth can be ascribed to genetic potential and nutritional experience, rather than other biologic factors. Let us first summarize how various forms of depressed size at birth are defined and then consider each of these interpretive issues.

Low birthweight, conventionally defined as birthweight less than or equal to 2500 grams (Behrman & Vaughan, 1983), may occur as a result of shortened gestation, depressed rate of growth during gestation, or both. Infants born after a shortened gestation, defined by the World Health Organization as a gestation lasting 37 weeks or less from the first day of the mother's last menstrual period (Behrman & Vaughn, 1983), are termed premature. Regardless of length of gestation, infants whose rate of growth during gestation is less than expected are termed intrauterine growth retarded (IUGR) or small for gestational age (SGA). The weights of some term or postterm (37–44 gestational weeks) intrauterine growth retarded infants may exceed 2500 grams, but still fall below the statistical norms for birthweight for their gestational age. Thus, while all low birthweight infants are either premature, intrauterine growth retarded, or both, not all intrauterine growth retarded infants are low birthweight.

As is the case for postnatal growth retardation, intrauterine growth retardation is assessed by comparing growth parameters not only to norms for gestational age, but also to one another. Intrauterine growth retardation is often classified according to Rohrer's ponderal index (weight in grams divided by the cube of the crown heel length × 100) (Miller & Hassanein, 1971). Infants with birthweights below the tenth percentile and with normal ponderal indices are considered to show symmetrical intrauterine growth retardation, with reduction in linear skeletal growth but normal soft tissue development. In contrast, infants with low ponderal indices, regardless of birthweight, are considered to show asymmetric intrauterine growth retardation, with reduced soft tissue mass but relatively normal linear skeletal growth (Miller & Jekel, 1985).

Now let us evaluate to what extent these heterogeneous forms of depressed size at birth can be interpreted as indicators of intrauterine malnutrition. Maternal nutritional status, expressed by mother's weight for height at conception and rate of weight gain during pregnancy, has been found to account for a significant proportion of the variance in both duration of gestation and rate of intrauterine growth, and thus in the variance in birthweight, in numerous investigations in both the developing world (Adair & Pollitt, 1985; Delgado et al., 1982) and the United States (Naeye, 1979). However, maternal nutrition is by no means the sole determinant of prematurity or intrauterine growth retardation. Both phenomena are final common pathways of multiple possible prenatal stresses to the maternal-fetal unit including, for example, bacterial and viral infections, maternal hypertension, and multiple gestation (Behrman & Vaughn, 1983). Symmetric intrauterine growth retardation may result from insults operating during the first 24 weeks of gestation, including intrauterine infections, chromosomal abnormalities, heavy maternal use during pregnancy of pharmacologically active substances such as cigarettes, alcohol, opiates, and anticonvulsants (Gaston, 1982; Villar et al., 1984), and chronic severe maternal malnutrition. In contrast, asymmetric intrauterine growth retardation is thought to indicate failure of adequate nutrient transfer to the fetus beginning in late pregnancy, occurring because of maternal hypertension, placental infarction, or maternal dietary intake inadequate to sustain rapid third trimester rate of fetal weight gain (Gaston, 1982).

Thus, although maternal nutrition is a critical determinant of birthweight, depressed birthweight in any individual case can rarely be unequivocably attributed to nutritional factors alone.

Neither prematurity nor symmetric intrauterine growth retardation can be used as a model for intrauterine malnutrition, since both may reflect multiple non-nutritional determinants. Of all low birthweight infants, only those born at term with asymmetic intrauterine growth retardation are usually considered as clear examples of intrauterine malnutrition. Therefore, in subsequent discussions, only asymmetric intrauterine growth retardation will be discussed as a model for the effects of intrauterine malnutrition on behavior.

Now let us turn from consideration of size at birth as a nutritional indicator to the second issue—the extent to which perinatal history influences the specificity of postpartum growth as a nutritional indicator.

The same non-nutritional factors that may depress the infant's size at birth also may confound the evaluation of postnatal growth as a purely nutritional indicator. First, in the early years of life, prematurity must be taken into account in evaluating whether postnatal growth is adequate or not. All premature infants will appear to have depressed growth in early life if evaluated by standards for growth at a given postnatal age based on the growth of term infants. To obviate this difficulty, investigators must correct growth assessment for the infant's gestational age by subtracting the number of weeks the child was premature from postnatal age at the time of measurement, and then evaluating the child's growth by standards appropriate for the corrected age rather than the postnatal age. Such corrections should be done for head circumference until 18 postnatal months, weight until 24 postnatal months, and height until 40 postnatal months, or significant differences in growth percentiles will be obtained (Brandt, 1979). Obviously, such a correction requires accurate information regarding the infant's gestational age, information which may not always be available in field settings.

A second difficulty arises in interpreting the growth of premature infants as indicative of nutritional experience even when growth parameters are corrected appropriately for gestational age. Premature infants are subject to myriad serious illnesses resulting from the immaturity of their various organ systems including, for example, respiratory distress and central nervous system derangement. These illnesses, by non-nutritional mechanisms such as hypoxia and deranged central nervous system regulation, may impair later growth. Moreover, for reasons that are poorly understood, it appears that even without major organ system dysfunction, children born with very low birthweight (less than 1501 grams) may have impaired potential for postnatal growth (Kimball, Ariagno, Stevenson, & Sunshine, 1982).

At the same time, premature infants are at increased risk compared to term infants for postnatal nutritional deprivation because they tend to have immature and ineffective feeding skills and be poor elicitors of parental nuturance (Pollitt, 1975). Thus in a research setting it may be impossible to ascertain whether growth deficits in a child who was born prematurely reflect nutritional deprivation or other medical constraints on growth, or both.

Like premature, very low birthweight infants, many symmetrically intrauterine growth retarded infants born at term show postnatal growth failure that cannot clearly be ascribed to malnutrition. Children with intrauterine infections or teratogen exposure, chromosomal abnormalities, or severe and early maternal malnutrition, often fail to catch up from the relative deficits in body size present at birth (Smith, 1977; Villar et al., 1984). In contrast, the growth failure in asymmetrically growth retarded infants appears to be nutritionally responsive, with evidence of accelerated growth rates if nutrition is adequate in the early postnatal months (Villar et al., 1984).

These possible constraints on postnatal growth imposed by prenatal factors, both nutritional and non-nutritional, have not been extensively considered in the literature on malnutrition and child behavior. The common assumption that attained growth predominantly reflects nutritional experience can best be supported when potential perinatal constraints on growth have been sought for and ruled out. Optimally, to assess such constraints, information is needed about the medical course of the pregnancy, as well as about the infant's size at birth, and the duration of gestation, obtained either from interview of the mother or from standardized examination of the infant immediately after birth (Dubowitz, Dubowitz, & Goldberg, 1970). Under field conditions such detailed perinatal information is often not available, so that the specificity of growth as a nutritional indicator is diminished.

In addition to genetic and perinatal constraints, postnatal growth may reflect other non-nutritional biologic stresses upon the child. Any major organ system dysfunctions such as, for example, cyanotic congenital heart disease, endocrine abnormalities, or cerebral palsy, ultimately will produce growth failure (Bithoney & Rathbun, 1983). Acute infectious illnesses may exert direct toxic effects on growth, independent of their nutritional effects, as described elsewhere in this chapter (Solomon, 1985).

In summary, nutrition is a critical but not a unique determinant of growth, either pre- or postnatally. In assessing attained growth, weight for height is a more clearly nutritional indicator than weight for age, but can only be used to assess recent nutritional experience. Height for age, which may reflect cumulative nutritional experience, is also influenced by a number of other non-nutritional constraints. The non-nutritional determinants of growth are known and can generally be determined by detailed medical histories and evaluation covering both pre- and postnatal life. When such information could be obtained, investigators of malnutrition and child behavior have sought to improve the specificity of growth as a nutritional indicator in their samples by excluding children with low birthweight or recognized chronic illness from the study (Galler et al., 1983a). In studies where there is not adequate information to exclude the non-nutritional causes of growth failure (e.g., Cravioto, Delicardie, & Birch, 1966), the developmental-behavioral correlates of growth failure cannot be unequivocably ascribed to malnutrition since many of the non-nutritional biologic constraints on growth such as prematurity, intrauterine viral infections, or cerebral palsy, are themselves known to be associated with adverse developmental sequelae independently of nutritional factors (Drillien, Thomson, & Burgoyne, 1980).

2.2.2 *Accuracy of measurement*

Unlike dietary intake, physical size is susceptible to highly accurate measurement under a variety of field conditions. Although it is impossible to abolish measurement error completely in any scientific endeavor, it has been repeatedly shown that observers with proper equipment can be trained to obtain anthropometric measurements within circumscribed limits of agreement on repeated measures of the same child by the same examiner, and on measurements of the same child at the same time by different examiners (Hubbard & Hubbard, 1985; Zerfas, Shorr, & Neumann, 1977). Weight measurement can be standardized to within 10 grams for children under 2 years and within 100 grams for children over 2 years (Hubbard & Hubbard, 1985). Agreement within .5 cm can be obtained for both recumbent length (in children under 2) and standing height (for children over 2). Head circumference measurements can be accurately obtained within limits of .2 cm upon repeated measurement. Sources of measurement error, usually related to improperly standardized equipment or failure to undress or position the child properly, are well described (Zerfas, Shor, & Neumann, 1977). Procedures for standardizing measurement techniques within and between observers are established (WHO, 1983). Thus, although interpretation of anthropometric data as a nutritional indicator may be a matter of controversy, accuracy of anthropometric measurement is, in general, not an issue in interpreting research in the field of malnutrition and child behavior, as long as trained personnel and standard equipment are used to obtain the measurements.

2.2.3 Reference standards

Until the later 1970s there was no single commonly accepted set of growth standards for the evaluation of children's growth in studies of malnutrition and child behavior or, for that matter, for any nutritional research (WHO, 1978). Some investigators (Graves, 1976; Hoorweg & Stanfeld, 1976) used local standards constructed for the region in which the study was conducted, while others (Galler et al., 1983a) applied the Boston or Iowa standards (Reed & Stuart, 1959; Stuart & Meredith, 1946), derived from middle- and upper-class American children of North European ancestry measured during the 1940s.

This lack of uniformity in growth standards created two difficulties in interpreting results of studies of malnutrition. First, the comparability of findings across studies could be questioned, since the standards used to define children as malnourished or well-nourished were not consistent. Second, there were systematic biases inherent in norms from different types of populations. In particular, norms derived from children in developing countries were criticized as being based on an unknown number of malnourished children. Thus they would be artifactually low and would lead to an underestimate of "true malnutrition" if used to evaluate study subjects. In contrast, norms established in Western countries were often criticized as overrepresenting obese children and thus causing an overestimation of "true malnutrition."

Further, two recurring controversies have influenced the selection of growth norms for nutritional research. The first was whether reference norms for growth should be descriptive or prescriptive—that is, whether norms should provide an objective description of commonly occurring growth patterns or should indicate optimal growth under ideal conditions. To evolve descriptive standards, measurements on a large and statistically representative sample would be necessary. Prescriptive norms presuppose a reference sample, which is not necessarily statistically representative but which excludes all children suffering from conditions that may cause less than optimal growth (Waterlow, 1984). The second controversy revolved around the issue of whether standards derived from children from one ethnic background (usually Caucasian) could be applied to children of other ethnic and racial backgrounds to identify malnutrition, or whether differences in growth identified in this way would reflect differences in genetic growth potential, rather than nutritional experience and other environmental factors.

Since 1978 (WHO, 1978), a consensus has emerged regarding these issues. The concept of prescriptive norms has been discarded in favor of statistically representative descriptive standards which, if used widely, permit meaningful comparisons between populations and within the same population at different times (Waterlow, 1984). Moreover, several investigators have marshalled evidence to demonstrate that norms derived from statistically representative samples in the developed world can appropriately be used to evaluate the nutritional well-being of children of difffering ethnic and racial backgrounds in both developed and developing countries, since in infancy and early childhood socioeconomic status (and thus, presumably, nutritional experience) is the primary determinant of growth. Thus, differences in growth between children of similar socioeconomic status but differing racial and ethnic backgrounds are small, while differences between children of the same ethnic background but varying degrees of social privilege are large (Graitcer & Gentry, 1981; Habicht et al., 1974).

The descriptive norms currently recommended for use in the evaluation of childhood growth for both clinical and research purposes are those established by the National Center for Health Statistics (NCHS) of the United States (Hamill et al., 1979; WHO, 1978). These standards are based on two reference samples. Norms for children under 2 years are based on data collected on 867 children followed by the Fels Research Institute in Yellow Springs, Ohio, while those for children 2 to 18 years are derived from a multistage stratified probability sample of over 20,000 American children of varying ethnic, socioeconomic, and geographic characteristics, measured according to standard techniques during sequential cycles of the Health Examination Surveys from 1963–1975. From these data, age and sex-specific smoothed percentile curves have been established for head circumference, weight for age, recumbent length for age (under 24 months), standing height for age (over 24 months), and weight for length and weight for height, which are considered age independent for prepubertal children (Hamill et al., 1979). Recent investigations of malnutrition and child behavior (e.g., Barrett & Radke-Yarrow, 1985; Graves, 1984) have classified subjects' nutritional status according to these NCHS norms.

In contrast to the consensus regarding the use of a single set of norms for evaluating post-natal growth, there is no consensus regarding norms for evaluating the rates of prenatal growth before term. There are a number of published standards which vary both with regard to the ethnic make-up of the populations studied and the altitude at which reference data were obtained (e.g., Usher et al., 1969; Lubchenco et al., 1963). Therefore, comparisons between studies using different norms must be drawn with caution.

2.2.4 *Cutoffs for evaluating physical growth*

Three statistical strategies have traditionally been employed to classify attained weight for age, weight for height, and height for age as indicators of the relative severity of malnutrition. The most common analyzes growth parameters as the percent of the standard (mean or median) of the reference population. Classifications have been derived empirically, based on risk of mortality for children whose growth parameters are at a given percent of standard. In contrast, the other two strategies, which rank subjects according to centile rankings or according to standard deviation units relative to a reference population, are based on statistical convention. Each approach has strengths and weaknesses which are outlined below.

In addition to classifying nutritional status by these criteria for attained size at a single point in time—for instance, at time of hospitalization or at time of recruitment into a study—some investigators have also categorized subjects' nutritional status by growth rate. Assessments of growth rate are based on sequential measurements of anthropometric parameters over time. The statistical criteria for evaluating growth rates are less uniformly established than those for attained size, and will be discussed only briefly.

The strategy of categorizing anthropometric parameters as percent of a reference mean or median was evolved by Gomez and his colleagues (Gomez et al., 1956) to classify malnourished children according to their relative risk of dying within 48 hours of admission to a hospital. According to these classic criteria, which have been widely used to identify subjects in studies of malnutrition and child behavior (e.g., Galler et al., 1983a; Graves, 1976; Hoorweg & Stanfeld, 1976), children are identified as normal or well-nourished when body weight is greater than 90% of the expected mean for age. Malnutrition is classified as mild or first degree when weight is 76 to 90% of the expected mean for age, moderate or second degree when weight is 61 to 75% of the expected mean for age, and severe or third degree when weight is less than 60% of the expected mean for age (Gomez et al., 1956). Waterlow (1972, 1974) subsequently extended and revised this approach to categorize height for age which, as described earlier, is considered an indicator of chronic malnutrition or stunting, and weight for height, the conventional indicator of acute malnutrition or wasting, as percents of reference medians. The actual cutoff percent of standard for classifying severity of malnutrition differs for each parameter. For example, according to Waterlow's scheme, chronic malnutrition is classified as mild if height is 90% to 95% of the median for age, moderate or second degree if height is 85 to 90% of the median, and severe if height is less than 85% of the median for age. However, weight for height deficits are considered mild or first degree at weights of 80 to 90% of the median weight for height, moderate or second degree at 70 to 80% of the median, and severe or third degree at less than 70% of the median for height.

In addition to the classic and widely used schemes of Gomez and Waterlow, several other authorities (Jelliffe, 1966; McLaren & Read, 1975; Wellcome Working Party, 1970) evolved classifications of the severity of malnutrition based on percent of standard of anthropometric measures in conjunction with physical signs such as edema. The class boundaries for mild, moderate, or severe malnutrition in these systems differ slightly from those used by Gomez and Waterlow, but the overall approach is similar.

The statistical approach of classifying anthropometric parameters as percent of a standard mean or median has two major advantages: percent of standard is easily calculated in field settings, and values derived can readily be manipulated statistically. However, this classic approach to categorizing malnutrition suffers also from important drawbacks. The cutoffs for dividing children into three categories were empirically derived on the basis of observed risk of mortality from malnutrition in hospitalized children.

When investigators apply these approaches to nonhospitalized populations, the thresholds at which risk of mortality increases seem to vary from setting to setting. For example, for weight for age, estimates have ranged from as high as weight at 80% of standard in one investigation (Kielman & McCord, 1978) to as low as 60% in another (Chen, Chowdhury, & Huffman, 1980).

Further, no potential sequelae of malnutrition except death were used to establish these classifications, so that their appropriateness for identifying relative risk of cognitive or behavioral effects of malnutrition can be questioned. In fact, even for mortality, the sensitivity and specificity of classifications based on percent of standard for weight or height are not impressive (Bairagi et al., 1985; Dietz, 1986).

In addition to these problems with the empiric justification of the Gomez and Waterlow criteria, there is a statistical difficulty as well. The method is age biased. With increasing age, a given percent of the median corresponds to a progressively greater deviation from the median (Mora, 1984), so that children of the same degree of malnutrition at different ages are not in fact comparable.

The second approach to classifying anthropometric parameters is to express the observed measurement as a percentile value within a reference population. This approach has been widely used in studies of failure-to-thrive (defined and reviewed in Chapter Three) in the developed world, where children are diagnosed by weights less than a given percentile ranging from the tenth (Oates, Peacock, & Forrest, 1985) to the third (Elmer, 1960). Percentiles have the same meaning at all ages, but have a serious disadvantage. The percentile approach cannot be used to discriminate degrees of deficit between children whose anthropometric measurements fall below the reference population distribution cutoff, usually the third percentile (Mora, 1984). Moreover, percentile rankings cannot be readily analyzed by parametric statistics.

Because of the difficulties inherent in both the percent of standard and the percentile methods of classifying anthropometric parameters, authorities currently recommend categorizing nutritional status according to standard deviation units, referred to as *z* scores (Mora, 1984; WHO, 1983). Such *z* scores have the same meaning at all ages, can be readily manipulated statistically, and can be used to quantify the relative standings of children whose measurements fall below the range of the reference population (Mora, 1984). The major disadvantage of this approach is that computation of *z* scores is somewhat cumbersome, particularly for weight which, unlike height and head circumference, is not normally distributed and requires logarithmic transformation. Moreover, it should be stressed that cutoffs based on standard deviations or *z* scores have been established on biometric criteria alone, and have not yet been validated in functional terms for mortality or any other potential outcome of malnutrition, including cognition and behavior.

Criteria for classifying nutritional status by growth velocity (change over time in height and/or weight) are less widely known than methods based on a single assessment of size. This is the case even though growth velocity is a more sensitive indicator of subtle derangements in nutritional status than is attained weight. Several approaches have been used. Vietze and colleagues (1980) classified children as failure to thrive if rate of weight gain over at least ten days was less than two-thirds of the fiftieth percentile rate. An alternate method described by Karlberg and colleagues (1980) classifies children as showing impaired growth if the z score of attained weight falls more than one standard deviation unit over three months. This approach has recently been refined to correct for regression to the mean (Heimendinger & Laird, 1983).

In summary, there is no ideal statistical strategy that is both biometrically standardized and empirically validated for classifying children's anthropometric measurement in studies of malnutrition and behavior. While classifications based on percent of reference mean or median have been most widely used, those based on standard deviation units are currently recommended.

Classification schemes based on growth velocity, which require accurate longitudinal data, have not yet been applied to the classification of children in studies of malnutrition and behavior, but may be useful in the future in studying effects of nutritional deprivation in its early stages.

2.2.5 Merits and limitations

In comparison to dietary intake data, anthropometric indicators of nutritional status have a number of advantages. Weight, height, and head circumference can be measured within narrow limits of known accuracy, evaluated by a single set of well-standardized international norms, and manipulated by well-established statistical techniques. However, unlike dietary intake, anthropometric indicators can never be interpreted as solely reflecting nutritional experience. Myriad non-nutritional factors, including genetics, perinatal history, and acute and chronic illnesses, must be evaluated and excluded before a child's physical size can be confidently ascribed to nutritional experience. Moreover, size at any one time of measurement reflects both past and current nutritional and non-nutritional factors, making inferences about timing of insult and observed effects difficult unless sequential measures are obtained. It is important to realize also that the anthropometric criteria traditionally used to classify children in studies of the effects of malnutrition upon cognition and behavior have been based either on empirically observed risk of death or on biometric convention. The anthropometric criteria that define risk for adverse behavioral or cognitive sequelae remain to be determined.

Finally, documentation of anthropometric deficits to identify malnutrition does not permit investigation of nutrition-induced changes in behavior which precede anthropometric deficits, although such behavioral changes may be of great functional significance. This issue is discussed further in Chapter Two.

3.0 PREVALENCE OF MALNUTRITION

3.1 Developing World

Malnutrition, whether defined by dietary intake or by anthropometric criteria, is a disease of poverty. In the developing world, acute and chronic limitations of the food supply place large proportions of child (and adult) populations in nutritional jeopardy. At the 1974 Food Conference it was estimated that only 38% of the 94 developing countries for which data were available had a total energy supply sufficient to meet the physiological requirements of their populations (Gurney, 1979). The number of people estimated to suffer from chronically insufficient diets varies with the criteria used to define inadequate intake, as Beaton (1983) has recently shown. However, whichever criteria are chosen, the calculated number of people at risk is staggering, with estimates for the early 1970s ranging from 599 to 808 million (Beaton, 1983).

When anthropometric criteria, usually weight for age, are applied to determine the prevalence of malnutrition among child populations in developing countries, the findings are overwhelming. For example, in Central America, it is estimated (Garfield & Rodriguez, 1985) that 24% of the children under 5 suffer from moderate or severe malnutrition by Gomez (weight for age) criteria. In this Central American population, protein-energy malnutrition is considered to be directly or indirectly responsible for 53% of the 100,000 deaths of children under 5 (Garfield & Rodriguez, 1985). Estimates for prevalence of moderate and severe malnutrition diagnosed by weight for age criteria in other areas of the developing world are comparable to those in Central America —19% in Latin America, 26% in Africa, and 31% in Asia (Bengoa, 1972, 1974).

Life-threatening moderate or severe malnutrition afflicts a substantial minority of the child population in the developing world. An even greater proportion of the same population suffers from mild malnutrition—with estimates as high as 40 to 50% (Barrett, Yarrow, & Klein, 1982; Srikantia, 1981; Viteri, 1981) by a number of investigators. When all categories of malnutrition among children under 5 are considered, the prevalence of malnutrition in children under 5 may be as high as 85% (Barrett et al., 1982; Garfield & Rodriguez, 1985) in many developing countries.

3.2 United States

In the United States, childhood malnutrition is also a disease of poverty, reflecting not the overall poverty of the nation, where food is, in fact, available in excess, but of individuals and regions within the country. There are four major surveys and several smaller investigations that examine the prevalence of undernutrition among pregnant and lactating women and young children in the United States. The four large surveys are the Preschool Nutrition Survey (PNS) and the Ten State Nutrition Survey (TSNS), both carried out between 1968 and 1970; the First National Health and Nutrition Examination Survey (N-HANES-I), carried out from 1971 to 1974, and the recently completed Second Health and Nutrition Examination Survey (N-HANES-II), which gathered data between 1976 and 1980. The results of the first three surveys have been summarized by Owen and Owen (1982). Livingston et al. (1975) summarized the results of the PNS and the TSNS.

Dietary recall data from these surveys indicate that a substantial proportion of low-income U.S. women of childbearing age have daily energy intakes below 75% of the RDA (approximately 2,000 calories a day for nonpregnant women, 2,300 calories a day for pregnant women, and 2,500 calories a day for lactating women). Prevalence of deficit is highest among women who are members of minority groups. For example, the N-HANES-I demonstrated that among women aged 18–44 whose income was below the poverty level, 18% of white women and 27% of black women were receiving less than 1,000 calories a day, while 48% of the white women and 57% of the black women were consuming less than 1,500 calories a day. This is consistent with the findings of the TSNS which classified subjects not by family income but by the income ratio of their state. The TSNS showed that approximately 65% of women in low-income ratio states and 45% of women in high-income ratio states received less than 1,800 calories a day during pregnancy. Caloric insufficiency of this magnitude is usually accompanied by deficiencies in other nutrients (Livingston et al., 1975). The evidence suggests, then, that a substantial number of low-income women in the United States do not obtain adequate calories during their childbearing years. Furthermore, energy intake among low-income women known to be pregnant is in many cases so low as to jeopardize fetal growth and development (Livingston et al., 1975), a finding corroborated by the excess rates of low birthweight and infant mortality found among offspring of low-income women, particularly women of minority origin (Wegman, 1985).

Low-income preschool children in the United States are, like their mothers, at risk from energy intakes well below age-specific RDAs. Livingston and colleagues (1975) calculated from the data of the TSNS and the PNS that 18% of infants under 1 year living below the poverty limit received less than 70% of the RDA for calories. Similarly, Owen and colleagues (1974) found that one-third of 1 to 4-year-old children from the lowest income quartile had substantially calorically deficient diets (less than 75 calories/kg/day, compared to an RDA of 100 calories/kg/day for 1- to 2-year-olds, and less than 60 calories/kg/day compared to an RDA of 80 calories/kg/day for 2- to 4-year-olds). The more recent N-HANES-II found that 20 to 25% of 1- to 5-year-old children assessed between 1976 and 1980 had diets that fell below these levels of caloric intake (N-HANES-II, 1983).

Two recent regional surveys reported anthropometric findings that corroborate the prevalence of nutritional deficiency among low-income preschoolers suggested by the dietary intake surveys, although absolute prevalence is lower when anthropometric rather than dietary intake criteria are used. The Massachusetts Nutrition Survey (MNS), which assessed 1,337 low-income children aged 6 months to 6 years presenting to primary care clinics in 1983, found that 10.5% of the children whose family incomes were below the poverty level had depressed heights for age, defined as below the fifth percentile (MNS, 1984). The Center for Disease Control summarized data from 32 states on children from birth to 4 years enrolled in federally sponsored health and nutrition programs and found that the prevalence of height for age below the fifth percentile increased steadily with age until by age 4 stunting was found in 16% of Hispanic and Native American children (Trowbridge, 1984). In both these surveys, while anthropometric evidence of acute malnutrition (wasting) was rare, evidence of chronic malnutrition (stunting) was common.

Although the data from the developing and developed world on the prevalence of malnutrition during prenatal life and early childhood are plagued with all the methodologic difficulties outlined in the foregoing sections, taken as a whole the results are consistent and appalling. An enormous proportion of the world's children from the moment of conception do not receive adequate energy to sustain normal growth and activity. While severe and moderate malnutrition, which pose the most immediate threats to life, are primarily confined to a significant minority of children in the developing world, chronic, less-obvious deprivation afflicts the majority of children in developing countries and a substantial minority of impoverished children in the United States where, as of 1983, 22% of all children lived in poverty (Baltimore Sun, May 23, 1985). As we shall show in subsequent chapters, there is now persuasive evidence that undernutrition in early life, even in its mildest chronic form, may exert significant adverse effects upon later cognitive and behavioral development.

Chapter Two: Methodological Issues in Research on Malnutrition and Behavior

1.0 OVERVIEW OF THE CHAPTER

The purpose of this chapter is to help place the research findings in perspective by examining the question of methodology in nutrition research. In the first section of this chapter we discuss the general issue of the relationship between theory, operationalization of variables, and research design, with special reference to the malnutrition research.

In the first chapter we discussed some of the methodological issues in the operationalizing and assessment of nutritional status. In the second section of this chapter we discuss the operationalization of behavior outcomes. We discuss different approaches to assessment of behavior, focusing on methodology for studying children's social and emotional functioning. In discussing different approaches to measuring behavior, we will address the topics of data collection versus data reduction, indicating the options available to the investigator for each operation; deciding on the appropriate unit of behavior for observation and analysis (i.e., the question of the 'level' of behavior analysis); and reliability and validity of observational measures. We will be introducing the concepts of intervention into natural situations as a means for studying child behavior, and discussing the problem of cross-cultural validity in malnutrition research.

In the third section of this chapter we discuss the question of construct validity of research design-the extent to which the research design fits the theoretical question. We discuss this problem in the context of the broader question of threats to the validity of research design. We have arranged this analysis using a taxonomy of problems in research design devised by Cook and Campbell (1979). Cook and Campbell describe four types of threats to the validity of a research design. The first, threats to statistical conclusion validity, are factors which reduce the power of statistical tests or which bias the tests in such a way that the true probability of making an incorrect inference from the data (i.e., of a relationship or no relationship between the variables of interest) differs from the presumed probability.

A second class of problems involves threats to internal validity. Internal validity is the degree of confidence one can have in concluding that a given independent variable does or does not have a causal relationship with a given dependent variable.

A third type of problem is external validity, defined as the degree to which we can have confidence in generalizing the results of a study to the population to which we wish to generalize.

Fourth, there is the issue of the construct validity of the theoretical inferences one makes, which Cook and Campbell refer to as "construct validity of putative causes and effects." The issue here is, given an empirical relationship between independent and dependent variables, does the relationship 'mean' what we think it means? Have we operationalized the variables in such a way that they correspond to our theoretical constructs and, in choosing the constructs to include in the design, have we adequately represented the theoretical issue which underlies the research?

In the fourth and final section of the chapter, we discuss the general classes of research design. We discuss the properties of each, including strengths and weaknesses, and applications to malnutrition research.

2.0 THE RELATIONSHIP BETWEEN THEORY, RESEARCH DESIGN, AND OPERATIONALIZATION OF VARIABLES: IMPLICATIONS FOR COMPARABILITY OF FINDINGS

In many areas of scientific study, the research design is dictated almost entirely by theoretical concerns. For example, in most areas of experimental psychology, the choice of a research design will be based on the extent to which the design logically addresses the theoretical issues of interest. Let us take, for example, the question of the effect of practice in paired-associate list learning on later learning of a new paired-associate list. If the concern was with the difference between short, frequent practice sessions compared to long and infrequent practice sessions, subjects would be assigned randomly to the different conditions, treatments would be implemented, and the results would be assessed. Issues such as number of subjects in each treatment and number of items in each test list would be addressed on the basis of psychometric considerations, but even these would have to be determined, ultimately, on theoretical grounds. For example, number of items will influence reliability, and both number of subjects and reliability will influence the sensitivity (power) of the statistical tests. How sensitive the test statistic has to be should depend on how large an effect one expects (i.e., theoretically), as well as on the 'costs' of not identifying an effect when a true effect exists. More clearly related to theory are such methodological concerns as homogeneity of the sample (a homogeneous sample may facilitate detection of effects, but may limit generalizability); item modality (aural, visual), content (picture, word), meaningfulness and concreteness; item presentation duration and interval (for practice and testing); number of practice trials; trial-to-trial interval and trial-to-test interval. All of these decisions will be made on the basis of what the experimenter thinks is important to know about how and what information is learned.

These rigorous conditions of research design frequently can be met in the experimental research. As a result, there often is a high degree of concordance between the investigator's experimental aims (i.e., as reflected in the experimental design) and the theoretical issue.

For reasons to be described in more detail, the researcher studying the behavioral effects of malnutrition often does not have the same opportunity to translate theoretical concerns directly into ideal research design. Because of the ethical issues regarding human deprivation, practical issues regarding the provision of food to needy persons, and logistical issues unavoidable in the conduct of field research, the independent variables often are defined in ways not dictated by theory but by circumstance. For example, the concern in a study might be with the effects of differences in the amount of a particular food supplement, but as a result of the particular nutritional needs of the study population, only levels of supplementation exceeding a certain threshold level are included. Or, more frequently, the concern may be with the effects of a particular nutrient deficiency on a behavior outcome, but because of the ethics involved in systematically depriving a population, an experimental deprivation design is not used, and a comparative retrospective approach is employed instead (see below). In both instances, nontheoretical reasons have determined the nature of the independent (i.e., presumed causal) variable. For example, in the first case, rather than the independent variable being differences in nutrient X within the range of A to H, it is differences in nutrient X within the range of D to H. The nature of the study (the theoretical question) has been influenced by the possibilities for implementing the design.

It is important to recognize that this type of influence—between design and theory—occurs in two directions. At times, the design is the 'given'; this indicates certain operationalizations of variables and thus specific possibilities for influences on theory. At other times, operationalization of the independent variable is the given. This makes certain designs possible or not possible, and the theoretical implications then are subject to the constraints of design and operationalization.

Let us consider these two types of influences in more detail. As noted above, in research on malnutrition and behavior, the researcher does not have the experimental psychologist's luxury of depending solely on theory to determine the specifics of research design. Most often (although not always) the scientific issue concerns the effect of a particular type of nutrient deprivation on subsequent functioning. The most compelling way to address this issue would be to deprive a sample of this particular nutrient (only) and to examine short- and/or long-term effects. For humanitarian reasons, such deprivation experiments generally are not performed. Thus, investigators often turn to one of two related questions, each suggesting a particular design. One question is, Do persons who have had the experience of this particular deprivation differ in those areas of behavior predicted to be affected by the nutrient deficit from persons who have not? A second question is,

Do persons who are known to receive suboptimal levels of this nutrient, and thus be at risk for particular developmental impairments, show improved performance on the outcomes of interest when they receive supplemental amounts of the limiting nutrient? The first question generally is addressed by a nonexperimental comparison design, often with data collected retrospectively. The second usually is addressed by an experimental or quasiexperimental intervention design. Both of these designs often are viewed as alternate approaches for addressing the original question of the effect of limiting nutrient X on behavior variable Y.

The studies resulting from such designs, however, will address different issues. In fact, the alternate designs may be more ecologically valid than the original design. The first type of alternative design compares persons with certain natural histories of nutrition and disease with persons with different histories. Presumably the only difference between the groups is presence or absence of a particular nutrition problem although, of course, this cannot be precisely the case, since the groups will differ, at the very least, with respect to whatever environmental or constitutional conditions resulted in the nutrient deficit. The design, despite the inherent flaw of confounding of nutritional history with other causative factors, is an ecologically strong one, since malnutrition *does* occur in the context of certain environmental conditions (and determinants) and does not occur in isolation (i.e., as represented by the original design). In the second design, persons with known natural histories of marginal malnutrition or histories placing them at risk for more serious malnutrition are the subjects, and they receive experimental nutritional rehabilitation. Again, there is an ecological validity in the sense that the subjects are experiencing conditions which are natural, not artificially imposed. Only in receiving a remedial intervention to improve function or prevent adverse consequences of their nutritional deficit are they experimentally manipulated. Even then, one is not imposing malnutrition on otherwise healthy subjects or drastically changing the environments of subjects so that the results of the intervention are no longer meaningful.

The alternate designs address different research issues. Rather than asking the question (i.e., the hypothetical question) of what would happen if we limited one nutrient only, we are asking the real-life questions of (a) how do people with a particular set of nutrition-related experiences differ from people without such experiences, and (b) what are the effects of intervening into the natural environments and nutritional regimens of persons at risk for malnutrition, with respect to later behavior and development?

The three designs not only address different practical issues. They often necessitate different operational definitions of variables, specifically, the independent variables. These different operationalizations mean that different theoretical issues are being addressed. This raises important questions about the comparability of findings from study to study.

Let us look at the three classes of design just discussed and the operations indicated.

First, consider the original design concerning the effects of a systematic deprivation of a nutrient on a person's behavior. Such studies generally are not performed with humans, although there have been a few studies of this kind performed on consenting adults (e.g., Keys et al., 1950). In such studies, the independent variable is operationalized as presence/absence of a particular nutrient, degree of deficit of a particular nutrient, or duration of deprivation (including a measure of subject age at onset independent of duration). The design focuses on the amount of the nutrient missing or present in diet, and there is no assumption concerning functional states. For example, changes in physical status or biochemistry accompanying the deficit are not part of the conceptualization of the independent variable (although they may be included as outcome measures).

The second study design involves a comparison of persons with and without a history of naturally occurring deficit of a particular nutrient. Here the presumption of deficit generally is based on hospitalization or clinical treatment. Now, the independent variable includes the total set of experiences resulting in the inference of history of clinical undernutrition, including hospitalization and medical treatment.

The third study design involves nutritional supplementation. The independent variable is presence or absence of nutritional supplement, and/or degree of supplementation, including duration of supplementation and a measure of onset of supplementation independent of duration.

Note that the first and third designs appear to use the same independent variable, albeit in reverse forms. The independent variables, however, are not the same. First, deprivation of a nutrient may involve different physical operations than addition. For example, supplementation studies often involve incidental extranutritional stimulation (e.g., interpersonal contact) for experimental subjects. But deprivation studies may necessitate more rigorous medical monitoring and thus even more interpersonal contact. Second, the deprivation of a nutrient cannot be expected to have consequences isomorphic with (although in the opposite direction to) addition of a supplement. There is little reason to believe that the physiological responses to a nutrient deprivation in healthy subjects should parallel precisely the physiological responses to supplementation in unhealthy subjects—even for the same nutrient—unless there is empirical evidence to indicate such a relationship.

We have, then, the three designs, all geared to the same central issue, but actually addressing different theoretical and practical issues. This results partly from the fact that the operations used to define the independent variables differ. We have, in one case, malnutrition defined as deficit of a specific nutrient (only). In the second case, malnutrition refers to a history of nutritional inadequacy as manifested by hospitalization (or other clinical treatment) for malnutrition. In the third case, malnutrition becomes the condition of needing nutritional supplementation but not receiving it (as opposed to needing supplementation and receiving it). These are the differences at the most general level—obviously, the specific nutritional questions addressed (regarding type, severity, and duration of undernutrition) may vary even more greatly.

What is most important is that, despite these concerns, investigators continue to compare studies using these different designs and operationalizations to make inferences about the same theoretical issues, without explicitly recognizing the differences in the methodologies and thus, in the theoretical questions which are addressed. Certainly, one can and should use each approach to obtain further understanding of possible mechanisms in the relationship between early malnutrition and later behavior. Differences in findings, however, often are vaguely attributed to differences in method without an appreciation of what this really implies. It is not just that the designs are different or that the variables are different. Because of the intertwining of operational definition of variable, research design, and theoretical ramifications, the very issues being addressed necessarily differ from study to study.

We noted earlier that the relationship can work in reverse. That is, it frequently is the case that the investigator first makes a decision about how to operationally define the independent variable. This implies a certain design and thus determines the research question that will be addressed. For example, a number of studies have chosen specifically to examine the effects of a nutritional supplement on behavior, with a focus on the supplement per se. In such studies, the concern is with the cost-effectiveness of a particular nutritional intervention for preventing otherwise adverse outcomes of undernutrition. When such a question is asked—i.e., about a particular type of nutritional supplementation—the scientific question is, In an endemically malnourished population, at risk for more serious malnutrition and/or developmental impairment, does provision of the limiting nutrient have a positive effect on the subjects with regard to the developmental outcomes thought to be in jeopardy? The theoretical issue in this case is not simply the relationship between -X and -Y (i.e., the limiting nutrient and behavioral impairment), because -X necessarily co-occurs with a number of other conditions (including precipitating conditions). Further, the theoretical issue could not be the functional relationship between -X and -Y, because X is added and not subtracted. In short, when one begins with the variable—so to speak—and formulates a design around it, a specific research question is generated. This question may or may not be the same as the underlying theoretical issue which generated the research project.

In fact, any decision about how to conceptualize and operationalize undernutrition has implications for research design. Certain designs follow from decisions about the nature of the independent variable. For example, when malnutrition is conceptualized in terms of growth impairment and impaired organ function resulting in hospitalization, studies generally involve comparisons of children with and without histories of hospitalization for severe malnutrition. When nutrient intake is studied as a causal influence on behavior, the designs generally take the form of experimental intervention studies in which amount of food consumed can be controlled. When individual differences in growth are assumed to be related to nutritional history and to have possible ramifications for later development, the design which results generally is correlational, considering the relationships between anthropometric measures and developmental impairment in free-living populations.

In short, conceptualization and operationalization of the independent variable and research design are inextricably related. The decision to conceptualize malnutrition in a particular way (i.e., with respect to ultimate cause, necessary criteria, or continuous vs. qualitative nature) may not always be deliberate and, in fact, may be dictated by other considerations (the possibilities for implementing a particular type of study, for example). But it invariably has implications for the type of research design which results. Similarly, every research design implies a certain conceptualization of the independent and also the dependent variables. To choose to do a supplementation study, for example, is to make certain assumptions about nutrient intake and behavior.

For these reasons, replication of findings across designs and variables strengthens our confidence in a functional relationship only insofar as similar processes can be assumed to underlie the empirical associations between independent and dependent variables. Further, failures to replicate can only weaken our confidence in the associations if we can assume that the same causal processes occur in the different experimental conditions. For example, we can describe the findings of retrospective comparison studies of children with histories of early protein-calorie malnutrition showing effects on later social development as consistent with studies showing effects of improved nutritional supplementation on the social behavior of children from endemically malnourished populations *if and only if* we are willing to hypothesize the same mediating processes linking together the nutritional and behavioral variables. If we assume that the processes mediating the effects of early undernutrition on later dysfunction are quite different from those mediating the relationship between supplementation and behavior improvement, it makes no sense to attempt to relate findings as 'replications' or 'nonreplications.' Under those conditions, we are in the strongest scientific position if we treat the studies as related, but addressing different theoretical and empirical issues, examining their findings in conjunction but not assuming that they should or should not concur.

In conclusion, conceptualization of the independent variable and research design dictate the nature of the theoretical issues that can be addressed by the investigator. One would hope that most studies begin with the question, What do I wish to know about nutrition and behavior? and then proceed, simultaneously, to the questions of conceptualization of independent and dependent variables and research design. It seems clear, from review of the research that has been done and the inferences made, that often the precise research issue has not been articulated carefully at the outset. Unless this is done, appropriate conclusions will not be drawn.

3.0 MEASUREMENT OF BEHAVIOR VARIABLES*

There are many different ways to measure behavioral variables, that is, to assign values to behaviors which indicate amount or quality of the characteristic of interest. We have found it useful to classify measurement procedures with respect to two operations: first, data collection—the manner in which the behavior sample is obtained; and second, data reduction—the manner in which meaning is assigned to behavior.

Given identification of a behavior variable (let us say sociability, aggression, or numerical reasoning), the nature of the empirical findings one obtains depends on the measurement operations—how the construct is defined, and then operationalized. We will not discuss how one goes about initially defining one's construct in this section (although we will discuss operationalization of the construct *and* construct validation). This is not a trivial dimension of research methodology; it is critical. It is not appropriate here, however, to discuss different conceptualizations of specific behaviors such as sociability or aggression (i.e., What do we *mean* by aggression?). These depend on theoretical orientations and on previous research support for the theoretical orientations. Suffice it to say that agreement on the part of investigators concerning

* Portions of sections 3.1 and 3.2 appear in Barrett, D.E. (1987) Undernutrition and child behaviour: What behaviours should we measure and how should we measure them? In J. Dobbing (Ed.), Early Nutrition and Later Achievement. London: Academic Press, pp. 89–119.

what is *meant* by a construct is a necessary first step in the measurement process.

Parenthetically, let us give an example from the research literature where a failure to conceptualize a construct carefully has led to disagreements about nutrition-behavior relationships. The animal research shows a clear pattern of relationships between malnutrition and emotionality, in the sense of disruption of normal adaptive responses to moderately stressful situations, manifested by behaviors which indicate autonomic disturbance and uncontrolled or highly constricted motor activity. In a rather severe attack on the research on malnutrition and emotionality, Levine and Wiener (1976b) noted the discrepant findings concerning activity level. Activity level, they observed, was commonly used as a measure of emotionality; that is, emotionally distressed animals purportedly became frantically active. They found conflicting findings, however, such that in some studies better-nourished animals were more active and in some, less. On the basis of this and other documentation, they questioned the validity of the commonly held supposition that malnutrition results in heightened emotionality.

The problem, of course, comes from the overly restrictive conceptualization of emotionality—that is, as necessarily resulting in heightened activity. It seems reasonable to assume that emotionality as defined above—disruption in adaptive responding resulting from autonomic stress—would be manifested in different ways in different situations, depending on the nature of the adaptive response. In some cases, heightened activity would be appropriate in the situation and a failure to explore or a tendency to freeze would indicate emotionality. In other situations, very high activity would be dysfunctional (i.e., interfering with task performance). In sum, it is critical that the construct of interest be defined carefully in studies of malnutrition and behavior. In Chapter Four we present an approach to the conceptualization and the assessment of social-emotional functioning in carrying out research on malnutrition and behavior.

On the following pages we list a number of approaches to data collection and data reduction (i.e., coding). It is important to note that some of the data collection procedures have certain coding methods which tend to 'go with' them. For example, when one uses naturalistic observation of work or play with time sampling, the behaviors observed are preserved on videotape or film, or are translated immediately into behavior codes by an observer who records the behaviors on paper or calculator. This documentation typically is used to generate frequency counts of discrete behaviors, may allow for coding of interactions (i.e., contingent responses), and may allow for scores on durations of behaviors. One also could use these data to generate global ratings of subjects, or code for presence/absence of behaviors. Or, to take another example, event sampling is used to generate detailed information about a low-frequency event, for example, child-to-child aggression. Often such data are recorded in a narrative description, and coding usually entails recording the sequences of the behavior acts (e.g., provocations, responses, retaliations). Conversely, one also could use the event sampling approach to arrive at broad judgments about the children in the sample (i.e., over and above discrete behaviors). Thus, the present separation of data collection from data reduction procedures is made in order to clarify the distinction between how the data are obtained and how they ultimately are assigned meaning, but should not be construed as indicating operational independence between the two stages of measurement.

We have described data collection methods in terms of six categories: naturalistic observation with time sampling, naturalistic observation with event sampling, experimental situations, naturalistic interventions (quasiexperimental situations), diary records, and standardized tests (also a type of experimental situation).

3.1 Methods of Data Collection

3.1.1 *Naturalistic observation with time sampling*

Naturalistic observation is an important method of collecting data in child development research. The essential features are that there are predetermined behavior categories, that behaviors are sampled over representative situations and over time, and that recording is done 'on the spot', whether mechanically or electronically. The strengths of the approach are the possibilities for generating individual scores for rates of behavior, the generalizability of observations to real-life situations, and the possibility for unobtru-

TABLE 2-1

Major strengths of five data collection methods

	TS	ES	EX	NI	DR
Replicability of assessments			*		
Estimates rates of occurrence in real-life situations	*				
Applicable for 'low-rate' behaviors		*			
Addresses hypothetical question regarding situational response			*		
Combines features of naturalistic observation (time sampling) and experimental situation				*	
Unobtrusiveness of assessment (with implications for representativeness of behavior sample)					*

Methods: TS: naturalistic observation with time sampling. ES: naturalistic observation with event sampling. EX: experimental situations. NI: naturalistic intervention. DR: diary record.

From Barrett, D.E. (1986). Behavior as an outcome in nutrition research. *Nutr. Rev. Supp.* 44: 224–236. Reprinted by permission of International Life Sciences Institute-Nutrition Foundation.

sive assessment. Other characteristics are shown in Table 2–1, which shows the major strengths of the five observation methods described here (i.e., the first five methods in the present listing). An application of this method to malnutrition research is given in Barrett, Radke-Yarrow, and Klein, 1982; see also Chapter Four of this book.

3.1.2 Naturalistic observation with event sampling

Often an investigator is interested in a particular behavior but the behavior occurs with such low frequency that it is difficult to observe reliably. Or, the investigator may wish to describe a certain type of episode in detail but may not wish to record all surrounding behaviors. In such cases, a fruitful approach is to observe until a particular type of episode begins to occur and then record the specifics of that episode. For example, in child development research, this approach has been useful in describing the nature of children's aggression and identifying dominance hierarchies in play groups (see Strayer & Strayer, 1976). In the malnutrition research, a version of this approach has been used by Nerlove et al. (1974) in describing the nature of children's day-to-day activities. In this study, on-the-spot, detailed recordings were made of the games and activities of Guatemalan village children and adults, and the episodes later were coded for the content and complexity of the behaviors. If the investigator is concerned with generating reliable rates of behaviors for the individuals in a study, this approach probably would not be chosen, and a more time-efficient approach—one which allowed the observer to record and make use of all the subject's behaviors which were accessible—would probably be used.

3.1.3 Experimental situations

Major theoretical breakthroughs in behavioral research have been made because investigators were able to arrange experimental situations which would allow the subjects to show responses to externally imposed manipulations. Much of what we now know about children's aggression (see Leibert & Baron, 1972), attachment behaviors (see Ainsworth et al., 1978) and prosocial proclivities (see Staub, 1978), as well as the potency of modeling situations (see Bandura, Ross & Ross, 1961) has been learned because researchers were ingenious in devising replicable situations which might elicit or make possible theoretically important responses.

In an experimental situation, the investigator has more control over the situation than in naturalistic observation. In fact, the experimental control and the possibility for replication of the observation is the major strength of experimental methodology. Replicability of the procedures does not imply reliability of measurements (see below) but does, to some degree, simplify the training of observers and, in this way, enhances the possibilities for interobserver reliability.

It is often pointed out that laboratory experiments lack external validity—that is, that the results of the assessments do not generalize to real-life situations (Babbie, 1975). It has been thoughtfully demonstrated, however (Mook, 1983), that external validity is not always a requirement for an assessment procedure. Very often, the major concern is what could happen under certain conditions (i.e., the conditions of the experimental intervention). At issue is not whether the response is representative of the subject's normal response; rather, it is whether a certain response might be elicited by a certain stimulus. Such an association may help to elucidate a causal association. (This, of course, is the rationale for the use of supplementation paradigms in studying the possible causal association between malnutrition and behavior impairment.) For example, in Liebert and Baron's study (1972) of the effect of television viewing on children's aggression, it was not important whether or not the type of aggression evidenced by the children in this situation generalized to aggression in natural settings. (The laboratory aggression involved having the children 'shock' an imagined person using a fake electronic apparatus.) What was important was that the study showed the possible effects of viewing television violence on the disinhibition of aggressive responses.

In the malnutrition research, experimental situations have been very useful in showing effects of malnutrition on animal behavior. Virtually all of the animal studies cited in Chapter Three—examining such outcomes as maze learning, passive and active avoidance responses, and emotional response to stress—have involved laboratory experiments. Most of the controlled manipulations in the human research have been embedded in naturalistic situations (see Naturalistic Interventions, below). An example of an experimental manipulation is the test of infant response to mild stress in the Bogota study (Mora et al., 1979), in which investigators looked at the infant's response to application of a cold disk on the abdomen. Vuori et al.'s study of visual habituation (1979) in the Bogota infants and Lester's studies of cardiac habituation of the orienting response (1975) in malnourished Guatemalan children also involved experimental assessments.

3.1.4 Naturalistic interventions

The strength of experimental testing is replicability. Because the assessment procedures are standardized, they can be administered in the same way from one occasion to the next. Experimental assessments, however, also involve a disruption of the subject's natural activity. This may not be a concern in terms of generalizability, since generalizability may not be an important consideration for the investigator, but it may be a concern in that the new situation into which the child is thrust may be so strange that a meaningful response is not generated. Thus, it may be useful to embed the situation in a natural context.

Examples of naturalistic intervention assessments are Rosenn et al.'s (1980) stranger approach assessment to study the correlates of nonorganic failure to thrive syndrome (NOFT), the structured situations (frustration task, new situation, impulse control game) used by Barrett and Radke-Yarrow (1985) to look at the effects of nutritional supplementation on social and emotional functioning in Guatemalan children (see Chapter Four of this volume), and Graves' mother-child interaction paradigm to study the effects of moderate malnutrition on maternal responsiveness and child exploration (Graves, 1976, 1978).

3.1.5 Diary records

Diary records involve daily recording of child behaviors, usually with a focus on developmental milestones or presence/absence of target (desired or undesired) behaviors. In malnutrition research, diary records have been kept to describe family food practices (i.e., to obtain measurements on the independent variable in the research) but generally have not been used to obtain behavioral records (i.e., to obtain measurements on dependent variables).*

The greatest strength of the diary record is unobtrusiveness of measurements. Recording is performed well after the behaviors of interest have occurred. There is no departure from routine and, in fact, the focus is on what occurs during the course of normal activity. Since a major concern of many nutrition investigators is whether or not children who have suffered from severe malnutrition early in life are capable of normal behavioral development, diary accounts of children's attainments of developmental milestones by parents and/or clinicians would appear to be a logical source of outcome data. Psychometric issues concerning parent report on child behavior are discussed by Radke-Yarrow, Campbell and Burton (1970).

* An exception is the use of daily recordings of school attendance as a behavioral outcome variable (see Stoch et al., 1982).

3.1.6 *Standardized tests*

Tests of abilities frequently are used in malnutrition research. IQ tests, tests of vocabulary, or specialized tests of such abilities as visual memory or discrimination learning involve, like other experimental situations, highly replicable procedures. The purpose is to discriminate individuals in terms of their responses to these standard situations. The experimental situations usually are referred to as tests when:

1. They focus on abilities;
2. They elicit responses which can be compared to typical responses of persons in the target population (these data often are published); and
3. Performance involves responses to discrete items or a single task presented in a *highly-standardized* fashion.

There is no dearth of data on the use of such measures in nutrition research. Standardized tests of intelligence and achievement are the most widely used measures of child behavioral development in the malnutrition research. See Pollitt and Thomson (1977) and Galler (1984) for reviews.

3.2 *Methods of Data Coding (Data Reduction)*

Once a sample of behavior has been obtained, there are a number of ways in which it can be assigned meaning. The decision about how to code behaviors will depend both on the particular variable being considered and at what level it is best conceptualized, and on the specific objectives of the research (i.e., What types of causal statements do the investigators hope to be able to make?). For example, it may suffice in some studies to rate each child in a global fashion for level of social-emotional adjustment, and to relate the independent variables to this rating. In another study, in which it is considered important to describe the specific dimensions of social and emotional functioning which could be affected by malnutrition, a more refined set of measures of outcome variables might be necessary. If the investigator felt it important to describe the effects of a nutritional variable on rates of specific behaviors, then frequencies of behaviors would have to be determined. If the concern is with describing the mother-child interactive processes affected by manipulations of nutrition, data in the form of contingencies of maternal response to child initiative (and the converse) might be desirable.

The question of how to assign meaning to behaviors is largely a question of the unit of analysis—that is, at what level of abstraction do we attach significance to behaviors? Is it worth coding a mother's kisses and touches, or should we make a more general rating of her 'affection'? Is it enough to code for the presence or absence of maternal imitation of the child, or should we count the number of imitations?

There also are the strictly psychometric considerations. It might seem reasonable to conceptualize 'affection' in terms of number of smiles and it might suit our research purposes to be able to describe the relationship between the independent variable and frequency of smiling. But it might also be difficult to measure this behavior reliably (i.e., in such a way that consistent individual differences were identified). In this case, it would be necessary to identify a measure of the construct which could be reliably measured. (We will deal with the questions of both reliability and validity below.)

Briefly, six different methods for reducing behavioral observations into countable units are: frequency counts, durations, ratings (of intensity, frequency, or quality), proportions of target response (for behaviors contingent on the occurence of another behavior), presence/absence of target behaviors, and clinical judgments or classifications.

Before describing these six categories of reducing data, it is useful to make the distinction between situation-general and situation-specific behavior assessments, and to discuss the different rationales for recording discrete behaviors (i.e., the observation approach) and making global ratings (i.e., the rating approach).

3.2.1 General approaches to assigning meaning to behavior: Observations and ratings

If we wish to assess the child's general behavioral characteristics (i.e., across situations), the most frequently used approaches are

(a) to collect and score time samples of behavior; and (b) to make global ratings of behavior.

Using the first method, the investigator observes the child's behavior in a variety of situations and, utilizing a systematic coding system, assigns scores to the behaviors of interest in each situation. The scores are aggregated across situations, thus providing summary scores (i.e., for each variable) which describe the child's central tendency with respect to the behavior of interest. Using the second method, the coder makes ratings of the child's behavior on the characteristic of interest, but does so on the basis of general knowledge of the child, rather than on the basis of how the child behaved in a specific situation. Ratings generally are made by persons who are familiar with the subjects (e.g., teachers or parents), although the research investigators may also be in a position to make ratings (e.g., in the study that we will describe in Chapter Four, the observers made general behavior ratings after all discrete behavior assessments were completed).

The strengths and weaknesses of the two approaches have been described by Cairns and Green (1979), who pointed out that the two methods have different objectives and yield different information. Briefly, the naturalistic observation approach is sensitive to contextual influences on behavior; that is, it will 'pick up' behaviors which are both situation-general and situation-specific. Thus, whenever we are attempting to arrive at scores that reflect the child's general behavioral characteristics or traits, this is a useful method only if we can sample widely across situations and time periods. Otherwise, the individual measurements will reflect short-term, situational influences (in which we presumably are not interested) and this will limit generalizability. The rating approach is, in some ways, less precise than the naturalistic assessment approach because it is based on a summary judgment about the child, made after the fact; that is, after the child's behavior (in a variety of situations and occasions) has occurred. It is highly influenced by a number of factors that do not influence the time-sampling observation; for example, the rater's personal concept of the variable of interest, the rater's selective memory, and the rater's bias with regard to the subject. Because it is not influenced by situational factors, since it represents a judgment which transcends time and place, the rating constitutes an assessment of the child's enduring (i.e., situation-general) behavior tendencies with respect to the variable of interest. It is for this reason that, given carefully trained raters, ratings made of children's behaviors often are more predictive than time-sampling observations of the child. This is because the rating is made with the understanding that situational or temporal variations in the child's behavior are to be overlooked, and scores assigned on the basis of what the rater sees as the child's general response tendencies. Thus, there will be a greater likelihood of showing a significant relationship with some other (presumably, also situation-general) personality characteristic.

When we are interested in describing a child's typical response to a particular type of situation, rather than to situations in general, either time-sampling observations or ratings may be used. In either case, the observations are made with the understanding that the variable of interest is a specified behavior under specified conditions, and relations between the independent variable(s) of interest and the measure of situational responding are sought.

In our research on malnutrition and behavior, we studied both classes of behavior: situation-general and situation-specific behavior. To assess the situation-general behaviors using the naturalistic observation procedure, we studied each child in a variety of situations and obtained an average frequency count of the behavior of interest per specified time interval (i.e., behavior rate) across situations. Subsequently we asked raters to make global assessments on a number of behavior dimensions based on their overall impressions of the child (i.e., after having observed the child in a variety of activities). To obtain scores on the situation-specific variables, we created structured situations for the subjects and observed their responses, making behavior ratings on the spot. Thus, although in general the assessments took the form of ratings rather than frequency counts, the approach was similar to the time-sampling assessments in that we recorded behaviors at the moment of observation and in the situation of interest.

3.2.2 Levels of behavior for analysis

3.2.2.1 Frequencies In many studies, the approach to deriving a score for the child's standing on a variable is to count the number of times the child performs an action. For example, in looking at the effects of malnutrition on sociability versus social withdrawal, one might count the number of times the child talks to other children in a given period of time. These behaviors can be converted to rates. Frequencies and rates can be reliable measures if (a) the behavior to be observed occurs with an adequate frequency, (b) there is agreement among observers about the criteria for what constitutes that behavior (e.g., Are all types of vocalizations 'talks'?), and (c) behaviors are observed over sufficient occasions and/or settings so that consistent individual differences will emerge. Epstein (1980) illustrates the importance of aggregation of individual measurements over occasions for reliability.

3.2.2.2 Durations When electronic recording of behaviors is available, durations of events often are used to indicate amount of a particular characteristic or trait. For example, duration of crying might be used as a measure of distress in response to a frustration or stress, or duration of play may be used as a measure of attention and/or persistence. When data are preserved on videotape and time is indicated on the tape, duration data are likely to be collected with high reliability, if the behaviors to be scored (e.g., cries, looks, vocalizes) are defined clearly. Under these conditions, latencies to respond—used, for example, in studying tolerance for frustration, impulse control, and vigilance—also may be coded with high reliability.

3.2.2.3 Ratings Ratings generally are construed as assignments of values to persons with respect to certain characteristics. The values may represent frequencies, intensities, or some other aspect of quality of the variable. In short, they assign the child to a place on a hypothetical continuum representing that variable. In the Barrett and Yarrow (1985) study of malnutrition and behavior, children were rated on scales of 1 to 5 on such variables as energy level, work orientation, and communication ability. In Galler et al.'s study in Barbados (1983b), children were rated by teachers on a number of different dimensions of personality, including attention and conduct. The strengths of the rating method were described earlier. It should be noted that ratings do not preserve the specifics of child-child interactions or child behaviors in specific situations. Thus, while they may be useful in differentiating individuals, they do not provide details about how a child behaved in a given situation.

3.2.2.4 Proportions of target responses Very often a behavior of interest is defined in terms of how a child responded to a particular stimulus demand. For example, tests are scored in this way. We look at the proportion of items to which the child responded correctly. With behavioral data, we often focus on the child's response to a particular behavioral antecedent, for example, how the child responds to a bid for attention or help, or to a provocation (see Barrett, 1979).

To code at this level of complexity, one must go beyond the child's behavior to include (in the behavior code) the provoking stimulus. When attempting to characterize individuals with respect to how they respond to particular situational cues, one arrives at scores indicating individual differences in proportions of responses having certain characteristics. Such proportion scores have to be treated very carefully in data analyses. They depend, for reliability, on reliable observations of both antecedents and consequences. Individual differences in scores on variables representing behaviors for which either the response occurs infrequently or the antecedent itself occurs infrequently are unlikely to have high reliability.

3.2.2.5 Presence/absence scores Frequently, children will be scored on a variable in terms of whether or not they exhibited a particular characteristic. For example, children may be scored for whether or not they were held back in school, had criminal records, or had psychiatric problems. Often such measurements can be made very reliably—particularly for variables such as the above, which can be agreed upon easily. In general, however, when a continuous dependent variable is assumed (e.g., for a construct like aggressiveness, helpfulness, assertiveness, or patience), ratings of children along the continuum typically are more reliable than forced-choice dichotomies. Barrett has shown (Barrett, in preparation) how rating scales for maternal behaviors during a teaching situation which include only Yes/No codings of the behaviors often may not capture the individual differences that can be obtained using global rating or discrete frequency approaches.

3.2.2.6 Clinical classifications Classifying individuals by making clinical judgments is the most frequently used method for describing the adjustment problems of adults receiving psychiatric or psychological treatment. This approach also is used to suggest methods of treatment and the likely effects of treatments. Terms such as schizophrenia, depression, and adjustment reaction are terms that most lay people, as well as scientists, recognize as clinical judgments, assigned to individuals based on observed behaviors, using techniques such as interviews, tests, and real-life observations. Such classifications often can be made reliably by trained clinicians (see DSM-III, 1982, pp. 468–469).

Recently, the classification approach has been used to describe individual differences among children in mother-infant attachments. The Strange Situation is an experimental paradigm devised by Mary Ainsworth (see Ainsworth, 1979) to study how children respond to separations from the parent, approaches of a stranger, and reunions with the parent, and how they use the parent as a secure base for play and exploration. This method yields classification of children into a group described as 'securely attached' or one of two groups described as 'insecurely attached.' It has been demonstrated (Waters, 1979) that observers can be trained to make such classifications reliably. The distinction between 'secure' and 'insecure' attachments has proved useful in identifying the sequaelae of early familial stresses and in predicting later child development outcomes (see Lewis et al., 1984).

3.3 Reliability and Validity of Observational Measures

The issues of reliability and validity are major concerns in all research. The reliability of a measure is the extent to which it elicits consistent individual differences. Reliability is a characteristic of a specific instrument, used in a particular way, in a specified population under specified conditions (i.e., in time and place). Reliability is estimated in a number of different ways. All methods have in common an evaluation of the instrument's abilty to consistently order individuals along a continuum representing the trait in question. For noncontinuous variables, methods of estimating reliability assess the consistency of the instrument in allocating individuals to categories (see Fleiss, 1981, Ch. 13). The methods most frequently used to obtain information on reliability involve testing a relatively large number of individuals on more than one occasion. An occasion is construed broadly to refer to circumstances of measurement. Thus, the individuals may be observed at different times, by different observers, or both. When the subjects are observed over time by the same observer, intraobserver measures of reliability are obtained (i.e., a test-retest reliability coefficient). When subjects are observed by different observers, interobserver coefficients are obtained. In both cases, the pair of assessments to be related to one another are considered parallel tests: measures of the same construct under similar conditions.

Another approach to measuring reliability is to measure the internal consistency of the test, or the degree to which the items or parts of the assessment relate to one another and to the whole. Split-half reliability coefficients are based on correlations between odd-numbered and even-numbered items in a test (or series of measurements) and are internal consistency estimates. Internal consistency estimates have been most frequently used for standardized, cognitive tests.

Technically, reliability is the proportion of variance in a measure which is true variance; that is, due to true individual differences. It is that proportion of variability in observed scores which is not due to error, where error refers to sources of individual differences which are random or not related to the amount of the trait in question (e.g., due to temporal factors). Different methods of estimating reliability differ in the sources of variance which they are forced to treat as error. For example, in test-retest reliability estimates, true changes in scores over time which affect the ordering of individual scores will contribute to error variance, whereas in interobserver reliability coefficients (based on measurements made at the same time), this will not be a source of variance in scores. A more detailed discussion of reliability—theoretical and applied—is in Magnussen, 1966.

Reliability of dependent measures depends on a number of factors and an essential factor is variability of observed scores in the population. Unless there is observed variability, there cannot be true variability and one cannot estimate the proportion of variance which is true. Lack of variability in behavioral measures can be a major problem in cross-cultural research. Often the measures are not meaningful to the study population and thus will not elicit a range of codable responses. We will address this issue in detail in a later section and will discuss an approach to identifying and resolving the measurement problems.

The reader interested in the general problem of reliability in behavioral research is referred to Johnson and Bolstad (1973) and Yarrow and Waxler (1979).

3.4 The Issue of Validity in Cross-cultural Research

The issue of construct validity is the most important consideration in determining the usefulness of a measure. The question of construct validity translates to: Does this procedure measure what we want it to measure? This is not an easy question to answer. Construct validation implies validation (or confirmation) of both the construct (i.e., the theoretical idea underlying the variable) and the theory (or theoretical framework) in which the construct is embedded. Put simply, construct validation is the procedure of determining whether the variable one has measured 'acts' the way one would expect it to on the basis of one's theoretical framework.

There are many approaches to construct validation. All are indirect and none alone gives a firm answer. At best we can say that the evidence does or does not support our contention that our measure of the construct is a valid one.

For example, if one is measuring emotionality in response to stress, one first would need to have some elaborated idea about what emotionality 'means', (i.e., what it relates to, does not relate to, is influenced by, influences, and, in the broadest sense, 'means' for a person's development). If, for example, one expected on the basis of one's theory that emotionality should relate to anxiety (assuming that the construct 'anxiety' were well defined and operationalized), but not to intelligence, one would expect that one's measure of emotionality would show a significant correlation with one variable (the measure of anxiety) but not the other (the measure of intelligence). If it were presupposed, by one's theory, that emotionality increased with uncertainty, then a failure of an uncertainty manipulation to increase scores on emotionality would raise questions about the validity of the emotionality measure.

Construct validation is a particular concern in cross-cultural research. Breslin et al. (1973) have demonstrated the use of factor analytic approaches to construct validation of behavioral measures in cross-cultural research and Leibel et al. (1982) have applied a factor-analytic approach to construct validation of cognitive measures in research on the effects of iron deficiency. Barrett (1982) showed the limitations of this approach for behavioral data.

Another approach to construct validation of a set of behavioral measures in cross-cultural research is to ask the question, Are differences in the mean levels of the behaviors within each population consistent with our expectations for how the behaviors should be distributed within a single population? A second question is, Are differences in the mean levels of the behaviors *between* the two groups consistent with the differences or similarities we would hypothesize based on our knowledge of the populations?

To illustrate, consider the data presented in Table 2–2, from Barrett, 1982. Here are shown differences in mean rates of behavior for a number of social and emotional behaviors and characteristics (measured using the time-sampling procedure described in Chapter Four) for two different populations: school children in rural Guatemala and school children in San Diego, California.

As shown in Table 2–2, the relative magnitudes of the rates of the different behaviors (i.e., within groups) are very similar. For example, in both groups, routine friendly interaction is much more frequent than either aggressive or prosocial behavior. With respect to affect, happy expressions are far more common than expressions of anger or sadness. And moderate activity level is far more frequent than low or high-activity level which, in turn, is more frequent than extremely high, uncontrolled activity. These are patterns that we might expect in any culture or population group.

TABLE 2-2

Comparison of Guatemala and San Diego children on selected interaction, affect, and activity variables

Variable	Guatemala (n = 138)		San Diego (n = 65)		Reliable differences[†]
	M*	SD	M*	SD	
Routine (friendly) interaction	1.99	0.76	1.72	0.57	> SD
Physical aggression	0.01	0.03	0.07	0.14	SD > G
Helps, shares	0.01	0.02	0.05	0.05	SD > G
Assertive (commands)	0.11	0.16	0.26	0.24	SD > G
No social interaction	1.83	0.83	0.26	0.15	G > SD
Happy	1.15	0.73	1.89	0.67	SD > G
Sad	0.004	0.03	0.08	0.21	SD > G
'Neutral' affect	2.60	0.61	1.70	0.62	G > SD
Angry	0.05	0.09	0.10	0.13	SD > G
Low activity level	0.49	0.65	0.45	0.46	
Moderate activity level	2.46	0.79	2.78	0.52	SD > G
High activity level	0.90	0.78	0.54	0.33	G > SD
Very high activity level	0.07	0.13	0.02	0.06	G > SD
No involvement with group activity	0.08	0.15	0.01	0.04	G > SD
Only watching	0.37	0.43	0.22	0.26	G > SD

* Mean frequency per 2-min interval.
† G. Guatemala sample; SD, San Diego sample

From Barrett, D.E. (1982). An approach to the conceptualization and assessment of social-emotional functioning in studying nutrition-behavior relationships. *Am. J. Clin. Nutr.* 35: 1222–1227. Reprinted by permission of American Society for Clinical Nutrition.

Most interestingly, the differences that we discern between populations—and all of the comparisons shown here are statistically significant—are consistent with the differences we would hypothesize based on our theoretical framework. For example, we hypothesized that undernutrition in early life would be related to less moderate (i.e., normal) activity level, less affectively charged peer behavior (e.g., aggression), less responsiveness to others, less positive affect, and less interest and involvement in activities at school age. Although the study was not undertaken primarily for the purpose of between-culture comparisons—rather, the focus was on relations between nutrition and behavior *within* populations—it was the case that the San Diego children were far more adequately nourished (i.e., in terms of mother and child nutritional history) than the Guatemalan children. When we examine the differences between mean behavior levels in the two groups, we see that in all between-group comparisons relevant to the hypotheses given above, the San Diego children scored higher on every variable we hypothesized to be negatively related to undernutrition and the Guatemalan children scored higher on every variable we expected to be positively related to undernutrition. Although these differences are no doubt due, to some degree, to other social and cultural differences between the groups, the findings are consistent with our theoretical orientation and increase our confidence in the validity of the behavior measures.

4.0 METHODOLOGICAL REQUIREMENTS FOR CONCEPTUALLY VALID RESEARCH STUDIES ON THE BEHAVIORAL EFFECTS OF MALNUTRITION*

4.1 Introduction

A number of papers and chapters have been addressed to methodological isues in research on the effects of malnutrition on human behavior (Elias, 1976; Tuddenham, 1974; Warren, 1973). Most have stressed the importance of designing studies that permit the investigator to infer with a high degree of confidence that statistical relationships between nutritional and behavioral variables reflect the influence on behavior of nutrition per se, and not the effects of one or more confounding (i.e., nonnutritional) variables. For example, Warren's review criticized a number of studies that used nonexperimental designs on the grounds that the variables representing nutritional status may have been correlated with social and environmental factors that themselves could have accounted for the statistical relationships. Tuddenham reviewed five classes of research design for studying the effects of malnutrition

* This section appears as Barrett, D.E. (1984) Methodological requirements for conceptually valid research studies on the behavioral effects of malnutrition. In: J.R. Galler (ed.), *Nutrition and Behavior*. New York: Plenum, pp. 9–36. Reprinted by permission of the publisher.

on behavior and described as adequate only those experimental designs that provided for random assignment to treatments and thus avoided the problem of selection bias. Klein and colleagues discussed several problems in the interpretation of retrospective studies of malnutrition and behavior, including possible confounding of onset and severity of malnutrition with duration of hospitalization, covariation of home environmental conditions with nutritional history, and difficulty in separating acute from chronic components of nutritional history. In short, the primary methodological concerns are summed up in the assertion by Elias (1976) that "the primary issue in the human studies is that of confounding of nutritional with non-nutritional factors" (p. 455).

It is interesting that the methodological critiques have focused on what Campbell and Stanley (1966) have called the internal validity of the research designs. That is, the concern generally has been: Has this study been designed in such a way that (a) variation in the outcome variable of interest can be unequivocally, although not necessarily uniquely, attributed to variation in the independent variable of interest (i.e., in the event that a statistically significant relationship is obtained), and (b) lack of relationship between independent and dependent variables is not an artifact of methodological or statistical features of the design or analysis that mask a true causal relationship? With respect to research on the effects of malnutrition on behavior, the internal validity of the design is the extent to which a significant relationship between malnutrition and behavior can be interpreted specifically as a nutrition effect and a lack of such relationship can be interpreted specifically as indicating no causal relationship between malnutrition and behavior.

Internal validity, however, is only one consideration in research design. There are several requirements that must be met if the research is to have heuristic value. A useful conceptual framework for evaluating the adequacy of a research design is Cook and Campbell's (1979) hierarchical classification of threats to the validity of a research design. There are four levels of evaluation.* The

* Cook and Campbell list the four classes of threats to the validity of a design in the order: statistical-conclusion validity, internal validity, construct validity of putative causes and effects, and external validity. I have reversed the order of the last two classes of validity for the purpose of clarity and logic of the presentation.

first is statistical-conclusion validity, the extent to which the data permit the investigator to conclude that a statistical relationship does or does not exist, regardless of what the relationship may 'mean.' The second is internal validity, defined above. Next, there is external validity, the extent to which the results may be generalized to different samples, settings, and time periods. Finally, there is construct validity of putative causes and effects, the degree to which our conceptualizations of the independent and dependent variables match our operationalizations (i.e., measurement operations) and therefore the extent to which the theoretical inferences we draw from the findings are empirically justified.

In this section I will discuss specific problems of research design and methodology relevant to the issue of the effects of malnutrition on children's behavior. I will focus on those problems that derive from inadequate preliminary conceptualization of the independent and dependent variables and insufficient attention to the limitations and generalizability of the interpretations of significant relationships. I then will discuss the practical implications of these issues: the probable difficulties in formulating studies that adequately address all the methodological concerns, and suggested guidelines and priorities for designing studies that will have theoretical value. It should be emphasized that many of the problems I will discuss are relevant to research studies in a wide range of fields, and not solely to research on malnutrition and behavior. I will give particular attention, however, to methodological issues that are particularly germane to this area of study and will cite examples to illustrate these concerns.

4.2 Statistical-conclusion Validity

Threats to statistical-conclusion validity include such problems as insufficient numbers of subjects, low reliability of assessments of independent or dependent variables or both, high sampling error due to heterogeneity of sample or lack of uniformity in administration of treatments or tests, and use of inappropriate statistical tests (i.e., procedures that require mathematical assumptions that are not met by the data). In short, any factors that reduce the power of the statistical tests or bias the outcomes such that the probability of rejecting the null hypothesis differs from the chosen probability level may be considered threats to the statistical-conclusion validity of the study.

Two of these problems assume particular significance in the research on malnutrition and behavior. First, there is heterogeneity of scores on the dependent variable due to unintentional fluctuations in the application of treatment. In experimental studies of the effects of nutritional supplementation on selected behavior outcomes, it often is difficult to make the treatments uniform for subjects in the same treatment condition. When the investigator lacks control over how supplements are administered or consumed, variation in the use of supplement will result (assuming the supplement has some influence on the criterion variable) in increased variability of response within the treatment group and thus reduced power of statistical tests. Two studies that faced this potential problem are the Bogota, Colombia study (Herrera et al., 1980) and the New York City supplementation study (Rush et al., 1980). In the Bogota study, food supplements consisting of dry skim milk, protein-enriched bread, and cooking oil were administered to families in which the children were at risk for malnutrition. For those groups receiving the supplements, supplements were provided weekly with no control for intrafamily distribution and no prescription regarding supplementation of or substitution for normal diet. Family use of supplementation was significantly affected by prestudy nutritional status. For mothers who had been consuming less than 1500 calories per day during pregnancy, supplementation did not significantly improve total daily caloric intake. Only for mothers who were consuming more than 1500 calories per day did the supplementation treatment significantly increase total energy intake. Thus, comparisons between supplementation and nonsupplementation conditions in this study must overcome an inflated error term within experimental conditions, at least insofar as individual differences in the criterion variables were sensitive to differences in nutrient intake. In the New York City project, supplements were administered to women during pregnancy. Subjects were at risk for delivering a low birthweight infant. There were two types of supplement, one a beverage containing 322 calories and 40 g of protein per can, and one a beverage containing 322 calories and 6 g of protein per can, provided daily. Although supplementation was intended to add to rather than substitute for diet calorie and protein intake, the supplement was used as a food substitute by both treatment groups, with 20 percent substitution of calories in the high-protein supplement group and 12 percent substitution of calories in the low-protein supplement group. Whether or not there was wide individual variability in the degree of substitution within groups cannot be determined from the published data. To the extent that there was, the sensitivity of statistical tests may have been compromised.

The second type of problem involves reliability of assessments. The issue of reliability of indicators of nutritional status or nutritional history has been discussed in detail by Habicht, Yarbrough and Klein (1974). In their paper, the authors illustrate how the reliability of a nutritional assessment sets an upper limit on the correlation that that variable can have with any other variable. (If the criterion measure is perfectly reliable, the maximum value of the validity coefficient—i.e., the maximum correlation of the predictor variable with the criterion variable—is limited by the reliability of the predictor variable; specifically, the maximum validity coefficient is the square root of the reliability; see Magnussen, 1966, p. 149.) They show, for example, that when repeated measurements for length yield test-retest correlations of 0.97, the maximum correlation of length with another variable is 0.98, only 2 percent less than the perfect correlation.

Test-retest correlations for other nutritional assessments or indicators, however, may be far lower. For example, in the data from the Guatemala Institute of Nutrition of Central America and Panama (INCAP) study that they report, test-retest reliability for home protein intake based on three-month recall surveys was only 0.51, resulting in a maximum correlation coefficient of 0.71, almost a 30 percent reduction in sensitivity. The importance of the reliability of nutritional assessments is well understood by nutrition investigators, and the point need not be belabored. It should be noted, however, that a number of factors may influence the estimate of reliability—not only such factors as the type of statistic computed (test-retest correlation, coefficient of internal consistency, coefficient of equivalence), the heterogeneity of the sample (Magnussen, 1966), and the age at which measurements are made (particularly for test-retest reliability coefficients), but also the setting and conditions of assessment. A case in point is the study by Lechtig et al. (1976) on the usefulness of the urea/creatinine (U/C) ratio as an indicator of protein ingestion. In a field study of Guatemalan preschool children, the test-retest correlation for U/C ratio was 0.96 when based on same-day same-urine-sample correlations, but only 0.50 when different urine samples were taken at one-day intervals. In contrast, intraindividual variability of this magnitude is not found when measurements are obtained under the laboratory conditions of a metabolic ward. The results of the study of Lechtig and coworkers show the importance of collecting reliability data on nutritional measures under the same conditions that will characterize the research.

It is in assessing the reliability of behavioral variables that the researcher investigating the behavioral effects of malnutrition may confront particular problems. Very often, the researcher must study the behavior of persons in a culture that is different from that in which the assessment procedures were originally developed. However, for an assessment procedure to be reliable, to elicit consistent individual differences in response, it must be meaningful to the study population. If it is not, one will not obtain varied, codable, and consistent—that is, reliable—individual differences. I will illustrate this by discussing a behavior assessment procedure that we used in our research in Guatemala (Barrett, 1984a; Barrett et al., 1982). Since this procedure was developed in the United States for a non-Hispanic population, we had to modify it to make it appropriate for the Guatemala sample.

One of the hypotheses in our study was that differences in the adequacy of pre- and postnatal nutritional intake would be related to the child's ability at school age to control motor impulses. We used a Simon Says-type game to examine alertness and impulse control. When we first devised the procedures, we used the usual instructions; all behaviors were modeled, and the children were directed to imitate the model's movement if the model said, "Simon says do this," and to not imitate if the model said only, "Do this." When we used the procedure with San Diego, California, schoolchildren (see Barrett et al., 1981), there was, as expected, a wide range of responses. When we tried the game with village children in Guatemala, however, we found that *all* children imitated *all* modeled actions, regardless of the verbal directions.

Thus, there was no variability of response. In thinking about this problem and how to rectify it, we considered two facts. First, the culture we were studying was more 'traditional' than the population in which we had developed the procedures; in particular, there was a greater deference to authority figures and more rigid constraints on child and adult role behavior. Second, and related to the first point, the children at this age level—ages 6–8 years—were being trained in their school classrooms to imitate adults' actions. Much of the activity in the classrooms involved repetition of the teacher's words and actions (singing, marching, repeating the alphabet), and the children probably were not accustomed to having an adult demonstrate a behavior and direct them *not* to perform the modeled behavior. Therefore, we modified the procedures. Instead of providing visual cues for the subjects, we directed them to respond differentially to verbal cues only. We reasoned that these cues would exert less 'pull' on the subjects. The new directions were: If the model said, "One-two-three-yes," the child should clap; if the model said, "One-two-three-no,' the child should refrain from clapping. The new directions were very effective. There were wide individual differences in responding and, as we report in Chapter Four (see also Barrett & Yarrow, 1985), we found significant relationships between nutritional history and the child's speed and accuracy of responding in this task.

4.3 Internal Validity

As indicated earlier, the major concern of methodologists evaluating the validity of research studies investigating the effects of malnutrition on behavior has been with the internal validity of the designs. Internal validity concerns the justifiability of causal inferences. In an internally valid design, we have a high degree of confidence in our conclusion that a given independent variable does or does not have a causal relationship with a given dependent variable. Our confidence in making such inferences is lowered when there are one or more alternative explanations for how the independent and dependent variables were related statistically or why they failed to relate.

Cook and Campbell (1979) list a number of threats to the

internal validity of a design. Many are well known to researchers in this field and require no definition: selection effects, statistical regression, history, maturation, subject mortality, and instrumentation. All of these effects are made less likely by the use of random assignment of subjects to treatments and the use of one or more appropriate control groups. Campbell and Stanley (1966) and Cook and Campbell (1979) describe a number of experimental and quasiexperimental designs that provide different amounts of protection against these various threats to internal validity.

There are four other types of occurrences that may jeopardize the internal validity of a design. Cook and Campbell have called these "diffusion or imitation of treatment," "compensatory equalization of treatment," "compensatory rivalry," and "resentful demoralization," and all are particularly likely when the experimental treatment is a large-scale intervention (e.g., a nutrition or educational intervention) that is widely perceived as bringing substantial benefits to the recipients. Diffusion or imitation of treatment means that the group that is not assigned the treatment that is perceived as the most beneficial receives the treatment informally —either by managing to share the treatment with the experimental group or by eliciting similar treatment from the investigators. Compensatory equalization means that the investigators depart from their planned administration of the treatments and nontreatments and attempt to give all groups similar experiences, so that no groups are deprived of a positive intervention. Compensatory rivalry occurs when a group receiving a less favorable treatment attempts to try harder—precisely because of their less-advantageous condition—when tested on the dependent variable. Resentful demoralization would occur if the group receiving the less-advantageous treatment tried *less* hard when tested on the behavioral measure than the group receiving the more favorable treatment. Diffusion or imitation of treatment, compensatory equalization, and compensatory rivalry reduce differences between experimental and control groups, while resentful demoralization increases them. All affect the internal validity of the design in that they reduce one's confidence in interpreting the relationship or lack of relationship between independent and dependent variables in causal terms.

All threats to the internal validity of a study are relevant to research on malnutrition and behavior. In the remainder of this section I will illustrate the operation of three of these effects in studies on the behavioral effects of malnutrition: selection, statistical regression, and maturation-history.

As Warren (1973) showed in his review, a number of studies that have compared children with a history of undernutrition with matched controls who were not malnourished appear to have been compromised by selection effects. Two notable studies are those by Stoch and Smythe (1963, 1967) in South Africa and Chase and Martin (1970) in the United States. Stoch and Smythe compared school-age children who had been severely malnourished in infancy with controls matched for age and sex on a standardized intelligence test. Although the nonmalnourished controls scored significantly higher on the intelligence measure, they also scored significantly higher on measures of social, economic, and educational status that were likely predictive of school-age intellectual test performance. Chase and Martin compared three- to four-year-old children who had been severely malnourished in the first year of life with a sample of matched controls on a standard developmental assessment. The nonmalnourished group scored significantly higher than the controls on the developmental assessment, but also scored higher on measures of the quality of the home environment. The authors themselves concluded that they could not separate the effects of malnutrition from those of "associated environmental influences." (p. 938)

The importance of statistical regression is illustrated by a study by Mora et al. (1974) that was a pilot study for the Bogota intervention study (Herrera et al., 1980), cited earlier. In the pilot study, 218 children between the ages of 2 and 5 years were assigned to a nutritional intervention in which the children received directly, food supplements containing 100 percent of the recommended daily allowances for protein and greater than two-thirds of the recommended allowances for calories. Of these children, 102 had been classified as malnourished prior to the intervention and 116 as well nourished, based on weight and height for age. A total of 76 malnourished children and 84 well-nourished children were assigned to a nonintervention control group. Children were tested on entry into the study and at a one-year follow-up on the Griffiths test of mental abilities. Both well-nourished groups showed slight decreases in Griffiths test scores, while the nonintervention malnourished group showed a slight increase. The results indicated regression-to-the-mean effects. It was by virtue of the significantly greater gain in the intervention malnourished group relative to the nonintervention malnourished group that the authors were able to conclude that nutritional supplementation had a significant and positive influence on cognitive development.

Maturation-history effects are changes in performance that are due to developmental and experiential changes in the subjects that can be presumed to have occurred regardless of whether or not there was an intervention. A special case of maturation-history is the cumulative deficit effect, described and documented by Jensen (1974, 1977). Cumulative deficit refers to a process whereby children from underprivileged populations—those characterized by poor educational history, impoverished physical environment, and poor health—show poorer functioning on tests of cognitive ability over time, relative to children from more privileged populations. According to the learning theory interpretation of Campbell and Frey (1970), such children are accumulating information and problem-solving skills at a lower rate than nondeprived children and thus are testing at lower and lower levels with repeated assessment. Since the child subjects of studies of malnutrition frequently are from populations in which formal learning opportunities are limited, such cumulative deficit effects are likely to occur. To the extent that they do, they present an obstacle that the environmental intervention must overcome.

Such a maturation-history effect is illustrated in a recent study by Zeskind and Ramey (1978, 1981). The investigators studied the effects of a cognitive and nutritional intervention program on the growth and development of prenatally undernourished infants. One group of subjects, one-half of whom were assigned to the intervention group and one-half to the control group, had been born with a low ponderal index (PI), a ratio of birthweight to birth length, which is an indicator of prenatal undernutrition. The other group, also randomly assigned to treatment and control conditions, was of normal PI. Children were assessed on the Bayley scales at 3 and 18 months and on the Stanford-Binet at 24 and 36 months. The pattern of results for the low-PI nonintervention children indicated a cumulative deficit with Bayley scores decreasing from 91 to 87 and Stanford-Binet scores from 78 to 71. In contrast, nonintervention normal-PI children showed a decline in Bayley performance from 3 to 18 months, but a slight increase in Stanford-Binet scores between 24 and 36 months. Intervention was associated with increases in the Bayley scores from 3 to 18 months and in the Stanford-Binet scores between 24 and 36 months for both groups. It appears that the effect of the intervention in the undernourished children was to help to reverse the cumulative effects of inadequate environmental stimulation and learning opportunities. The results suggest that in studies considering the effects of nutritional interventions on children's behavior and development, it may be important, in evaluating the improvements due to treatment, to consider whether one might expect decrements in performance over time without intervention. In short, the baselinc against which one evaluates the usefulness of the intervention may have to be adjusted to take into account the effects of cumulative deficit. It should be noted that the direction of effects for the maturation-history effect described above is opposite to that for regression to the mean. It is conceivable that both factors may be operative in a single study and to some degree cancel each other out.

4.4 External Validity

When an investigator has identified a relationship between an independent and a dependent variable and infers that the independent variable exerts a causal influence on the dependent variable, he or she also must ask the question, To what extent can I expect this relationship to generalize to different populations, different settings (for treatment or testing, or both), and different epochs? While the last question can be addressed directly by including different cohorts in the study (see Herrera et al., 1980; McKay et al., 1978), the issue of the generalizability across population groups or experimental conditions or both often is left as a matter of speculation.

In studies of the effects of malnutrition on behavioral development, the issue of generalizability across populations is extremely important. There are both theoretical grounds and empirical data to suggest that the relationship between individual differences in nutritional history and behavioral development will be highly dependent on the general level of nutritional status of the study population (as well as on variability in nutritional status) and on the adequacy of the environment, specifically, the opportunities for interaction with the social and nonsocial environment and thus for learning.

Lloyd-Still (1976) has argued that malnutrition may have its strongest effects on mental development when overall environmental stimulation—physical, linguistic, perceptual—also is limited. His argument is based on evidence that differences in cognitive development between children with a history of severe malnutrition in early life and those without such a history are consistently significant only when the children from both the index and the comparison groups have experienced impoverished social environments. Lloyd-Still's position is consistent with the findings of Frankova (1974) and Levitsky and Barnes (1972), who found in their work with animals that provision of a highly stimulating environment could offset the behavioral effects of malnutrition. Further support for the proposed interaction of malnutrition and environmental deprivation is found in the neurophysiological literature showing effects of both malnutrition and environmental stimulation or deprivation on brain structure and biochemistry (Greenough, 1975; Im et al., 1976; Rosenzweig & Bennett, 1972). With respect to human studies, the cognitive developmental theory of Piaget (1970) provides an interpretation of these interaction effects. According to Piaget's theory, changes in cognitive structure or the person's mental capacities occur as a result of the person's interaction with the environment, which forces him/her to accommodate to new experiences. Individual differences in rates of development are due to genetically determined individual differences in learning abilities (i.e., the ability to bring about changes in cognitive structure through interaction with the environment) and gross differences in the quantity and quality of individuals' interactions (i.e., experiences) with the environment. From a cognitive-developmental view, one would expect that large changes in environmental stimulation (including stimulation mediated by nutritional improvements) might have their greatest effects when the prior level of environmental stimulation is lowest.

A number of studies on the effects of malnutrition on children's behavior are consistent with this interpretation and indicate important treatment-population interactions. Klein (1979) studied the effects of protein-calorie supplementation administered prenatally and in the first postnatal years on children's cognitive abilities at preschool age. Significant supplementation effects were identified only for subjects whose families scored lowest on a composite measure of social-environmental status. Barrett et al. (1982) examined the effects of pre- and postnatal caloric supplementation on the school-age behavioral functioning of Guatemalan children selected from the population of children studied by Klein. While nutritional supplementation showed significant relationships with measures of social behavior and emotional characteristics across villages, nutritional supplementation was related to cognitive functioning only in the village in which the absolute level of supplemental caloric intake was lowest. That is, only at the lowest level of energy intake were differences in caloric supplementation predictive of differences in intellectual functioning. In two studies cited by Lloyd-Still (1976), school-age children from middle-class populations who had been hospitalized for neonatal disorders and had been severely malnourished during their hospitalizations were compared with previously hospitalized but nonmalnourished controls on measures of intelligence (Lloyd-Still et al., 1974; Valman, 1974). There were no differences between index and control children in either study. In contrast, in nine studies in which subjects were from lower socioeconomic classes, significant differences in cognitive development between children with a history of malnutrition and children without a history of malnutrition were identified (see Lloyd-Still, 1976, p. 119). These findings suggest that in studies investigating the effects of malnutrition on later behavior, the adequacy of both the nutritional and the nonnutritional environment may have a significant influence on the outcome of the research.

Generalizability of relationships across conditions of treatment or testing or both is a conceptually difficult issue. To the extent that a treatment has a different effect in different conditions of administration, we may assume that it does so because the setting or conditions of administration influences the very nature of the treatment (i.e., its impact on the recipient). But if this is the case, if we are talking about the conditions of administration *changing the nature* of the treatment, it seems more appropriate to view such influences not as threats to the external validity of the findings, but, rather, as threats to the construct validity of our interpretations. The same can be said for treatment-setting interactions with respect to conditions of assessment of criterion variables. It frequently is the case in behavioral research that measurement of a behavior variable obtained under one set of conditions is very different from measurement of the "same behavior" under other conditions. For example, frequencies of children's prosocial behaviors in one situation (e.g., a helping situation) may be very different from frequencies of prosocial behaviors in a different situation (e.g., a sharing situation), and the two sets of behaviors may have very different correlates (see Payne, 1980). Rather than view this problem as one of external validity or generalizability of findings, it seems more useful, heuristically, to treat the two sets of assessments as measurements of different behaviors, i.e., prosocial behavior in a helping situation versus prosocial behavior in a sharing situation. For these reasons, when we consider the important issue of whether a nutrition-behavior relationship is replicable if we vary the conditions of administration of a treatment or the procedure for measuring our outcome variable, we may do so in the context of the broader question: Are the relationships we have identified between our measures of malnutrition and behavior telling us what we wish to know about the relationship between malnutrition and behavior, as we have conceptually defined these constructs? It is to this issue—the construct validity of our interpretations of relationships between nutrition and behavior—that we now turn.

4.5 *Construct Validity of Putative Causes and Effects*

The question of construct validity is really a question of the conceptual validity of our intepretations of findings. We are asking, here, both about the adequacy of our operationalizations of variables, that is, whether they correspond to our constructs, and about the appropriateness of our inferences about the relationships between variables, whether the processes that have brought about an empirical relationship are the same as those that we have hypothesized to account for the relationship.

Cook and Campbell (1979) identified a number of specific threats to the construct validity of a research study. These may be grouped into four general categories. The first problem involves inadequate conceptualization of the independent or dependent variables, leading to too narrow, broad, or otherwise inappropriate operationalizations of the variables. Treated as separate problems but actually relating to the same general issue are what Cook and Campbell call monooperation bias and monomethod bias. Monooperation bias occurs when the investigator measures only one aspect of the dependent variable or operationalizes the independent variable in a restricted way. An example of monooperation bias is examining effects of malnutrition on general cognitive development, but including as a dependent measure only a test of verbal ability. The risk, of course, is that the investigator will draw an inference about the effects of malnutrition on intellectual development solely on the basis of findings about verbal abilities. Monomethod bias occurs when an investigator examines the relationship between independent and dependent variables, but studies the dependent variable using only one form of assessment; for example, responses to structured testing situations, but not behavior in natural situations. If we consider this problem as a failure to consider the target behavior in its entirety—that is, as a failure to comprehensively operationalize the construct—we have essentially another type of monooperation bias.

A second class of problem involves the interaction of treatment with prior testing and the interaction of different treatments. Both of these are threats to construct validity in that we may interpret a relationship as resulting solely from the influence of an independent variable, whereas in reality, the power of the independent variable to influence the dependent variable was affected by a prior measurement or a prior intervention. The first problem is an issue whenever the behavior being considered is psychological test performance. Testing effects and interactions are very likely to occur in mental testing, particularly of children who are relatively unfamiliar with test procedures (see Zigler & Butterfield, 1968). The second problem is common in field studies in which an intervention has been preceded by some preintervention familiarization process or informal intervention. In both cases, the failure to consider the interaction of the intervention with the preceding event or events may lead the investigator to interpret the identified relationships in causal terms that are not consonant with the empirical facts of the study.

A third type of threat to construct validity involves subject or investigator biases. Cook and Campbell refer to the interaction of treatments with hypothesis-guessing (a subject bias), evaluation apprehension (a subject bias), and experimenter expectancies (an investigator bias). In the first case, the relationship between independent and dependent variables is due not only to the putative effects of the treatment variable but also to the effect of the subject's knowing the hypothesis of the study and thus what behavior is desired or expected. A second type of interaction occurs when the subject behaves in the way that he/she does in an assessment as a result not only of the intervention but also of his/her concerns about how he/she should behave in the test situation. Evaluation apprehension effects are common when the test situation involves behaviors that could be viewed as socially inappropriate or questionable, such as aggressive or emotional responses to a situation. The third type of problem occurs when the subject's criterion performance is affected not only by his/her standing on the independent variable but also by the fact that the investigator was not blind to the subject's standing on the independent variable and knew how he/she "should" behave. While these problems—hypothesis-guessing, evaluation apprehension, and experimenter expectancies—each also could act as a threat to internal validity (i.e., operate independently of treatment effects), they are more likely to interact with treatments in influencing outcome.

Finally, and particularly important with respect to research on malnutrition and behavior, are those threats that involve the investigator's failure to fully consider the theoretical implications of the research and, specifically, the processes whereby the independent variable might influence the dependent variables. Such failures may lead the investigator to draw inappropriate conclusions about the relationships between the theoretical constructs. Two threats to construct validity fall into this category. One is what Cook and Campbell have called confounding construct and level of construct. This occurs when the investigator examines the operation of the independent variable over a specific range and assumes that the relationship holds at some other (usually broader) range. This is a common problem when the independent variable involves dosages (as in drug or nutrition studies) or learning trials. It may be that when other ranges of the independent variable are considered, the relationship between independent and dependent variables is different than was originally thought. For example, a relationship that was linear at one level of the independent variable may be curvilinear over a larger range. In this regard, Horton (1982) found that the effects of carbohydrate and protein restriction on the endurance of obese adults were different after one and six weeks of treatment. While one week of restriction resulted in decrements in performance on an endurance task of moderate intensity, six weeks of restriction resulted in improved performance, apparently due to the subjects' abilities to use the prolonged period of restriction to adapt to the lowered intake and increase the utilization of lipids (Horton, 1982, p. 1233). If the data had been analyzed using a correlational approach, duration of dietary restriction and endurance would have been negatively correlated for the period 0–1 week, but positively correlated for the period 1–6 weeks, and the two variables would have shown a curvilinear relationship over the duration of the study.

The second problem is restricted generalizability across constructs. This occurs when the investigator fails to test for relationships between variables that can help confirm or disconfirm his/her hypothesis, specifically, when he/she fails to include in the hypothesized network of relationships between variables constructs that should be included by virtue of their (a) importance to the theory and (b) probable sensitivity to the independent variable. This is a common problem in nutrition research and will be illustrated below.

In this section I will discuss what are perhaps the most important threats to the construct validity of research on malnutrition and behavior. First is inadequate conceptualization of the treatment variable, leading to interpretations of treatment effects that misconstrue the true nature of the effects. Second is inadequate conceptualization of the behavioral measures, seriously impairing the investigator's ability to make inferences about the theoretical relationships being investigated. Third is confounding of the construct with levels of the construct—the failure to specify, in designing and implementing the research, the levels of the independent variable that are being investigated, whether they represent randomly sampled or fixed levels, and why they, out of all possible levels, were chosen (if they were not randomly chosen). Finally, there is the problem of restricted generalizability across constructs. I will illustrate these problems, providing examples from published research on the behavioral effects of malnutrition.

The problem of inadequate preoperational conceptualization of the independent variable is most evident when the independent variable is a nutritional manipulation, such as nutritional supplementation. In such research, the treatment necessarily will entail more than simply the nutritional intervention. There will be interpersonal contact between researchers and subjects, provision of information and other cognitive stimulation, and health supports such as medical checkups or monitoring of nutrient intake. Under such circumstances, it is important that the investigator, prior to the beginning of the study, try to anticipate the role that such factors will play in the intervention and attempt to build them into his/her conceptualization of 'treatment.' The investigator may have to reconsider the hypothesized relationships between independent and dependent variables in light of this reformulation of the independent variable. Having defined carefully the independent variable with attention to the aforementioned considerations, the investigator now can attempt to match the operationalization of the variable—that is, the implementation of strategy—with his/her more fully elaborated conceptualization. Only when such steps are taken will the investigator be in a strong position to hypothesize and test for empirical relationships between variables and draw theoretical inferences consonant with the data.

What frequently occurs in conducting field research is that this type of consideration of the 'meaning' of the independent variable is not made, for reasons of time, convenience, or lack of awareness of its importance. As a result, the investigators may find, in attempting to implement the study, that they are exploring different relationships, testing different hypotheses, than they had anticipated. The nutritional treatment is no longer a nutritional treatment as such, but, rather, an amalgam of various types of stimulation—social, nutritional, cognitive, and medical. Pollitt and Thomson (1977) show in their review of the literature on malnutrition and behavior how this problem has made difficult the interpretation of the results of the major intervention studies (p. 295).

Another problem related to the requirement for adequate pre-operational conceptualization of the independent variable is illustrated by the experiences of the investigators in the INCAP longitudinal study, carried out in rural Guatemala (Klein, 1979). One of the interpretive issues in analyzing the INCAP data has involved understanding the nature of the dramatic supplemental calorie effects on growth and development (see Barrett et al., 1982; Klein, 1979; Klein et al., 1977). The INCAP study originally was designed to compare the effects of two nutritional supplements, a protein-calorie supplement and a calorie-only supplement, in promoting the growth and behavioral development of children from an endemically malnourished population. The calorie-only supplement contained no protein and approximately one-third the calories of the protein-calorie supplement. As the study progressed, it became clear that the most consistent predictor of postnatal physical growth and psychological development, of the relevant diet and supplement nutrition measures, was caloric supplementation, both pre- and postnatal. Supplemental calorie intake was a significant predictor of birthweight (Lechtig et al., 1975), height and weight at preschool age (Klein et al., 1977), and social and emotional characteristics at school age (Barrett et al., 1982). Because of the perfect confounding of supplemental calories with supplemental protein in the protein-calorie-supplemented children, one cannot separate the effects of caloric supplementation from protein supplementation unless one studies only those subjects receiving the calorie-only supplement and examines the implications of individual differences in supplemental caloric intake in that group. To do so, however, severely limits the conclusions one can draw about the significance of supplemental calories because the absolute *level* of supplement calories was lower in the calorie-only subjects than in the protein-calorie subjects. A second problem involves the confounding of calorie-only versus protein-calorie supplement treatment with village membership (i.e., village differences). Thus, if different effects of calorie supplementation are found for children receiving supplemental calories with supplemental protein compared to children receiving supplemental calories without supplemental protein (and such differences have been found; see Barrett et al., 1982), such differences could be due to (a) presence/absence of supplemental protein; (b) absolute level of caloric supplementation, or (c) village membership. It should be noted with regard to the last point that although the study villages were originally matched at the start of the study on the basis of a number of socioeconomic variables including nutritional status (see Klein, 1979), by the time the children were born, mothers from the different villages differed significantly in total prenatal caloric intake. A final interpretive problem arises because all children received supplementation beginning prior to birth. In a number of papers, INCAP researchers have attempted to investigate the importance of the timing of supplementation for later development (Barrett et al., 1982; Klein et al., 1977). The limitation of such analyses is that one can only draw conclusions about the timing of supplementation—or, more precisely, the differential importance of relatively high or low supplementation at different periods of development—for a population supplemented continuously beginning prior to birth. No statement is possible about the effects of supplementation beginning at a particular postnatal period, since no groups of subjects were staggered with respect to the initiation of the treatment. In sum, the INCAP study, while it has made a highly significant contribution to our understanding of the importance of chronic energy deficit for later behavior and development, is nonetheless limited in the theoretical contributions that it can make, since the nutritional variable that became the primary (i.e., most consistently used and most consistently significant) predictor of the physical and behavioral outcomes was not identified as such at the outset and therefore could not be completely adequately conceptualized and operationalized.

A second major issue involves the importance of careful, theoretically based conceptualizations of the dependent behavioral variables. I have addressed this issue at length elsewhere (Barrett, 1982) and will present here only the central argument.

In studying the question of how a history of malnutrition influences children's behavior, investigators frequently have chosen their dependent variables on the basis of nontheoretical considerations. For example, it frequently has been noted (Brozek, 1978; Ricciuti, 1981) that the majority of studies on the effects of malnutrition on cognitive development have conceptualized cognitive functioning in terms of general intelligence as defined by performance on standardized IQ tests. The problem with such a global and atheoretical conceptualization is not that one necessarily will fail to identify significant nutrition-behavior relationships; indeed, most such studies have shown that malnutrition is related to poor cognitive test performance (for a review, see Pollitt & Thomson, 1977). Rather, the problem is that unless dependent variables are explicitly theoretically derived and operationalized, the results of the research are unlikely to clarify the processes whereby malnutrition may influence later behavior. A notable exception in the area of cognitive research is the recent study by Leibel et al. (1982), which examined the effects of iron deficiency on specific attention and learning capabilities that have been linked theoretically to iron deficiency and empirically related to one another.

Several studies have attempted to assess the effects of malnutrition on children's social and emotional development. In a number of studies in this area, conceptualization of behavior variables has been weak. The reason may be that in most studies, the primary focus has been on the effects of malnutrition on intellectual functioning, with social and emotional functioning a secondary consideration. Whether or not this is the case, dependent behavior variables in the noncognitive domains have tended to be defined only in very broad terms, for example, social maturity or school behavior problems, and have been operationalized accordingly, that is, in a global fashion (see, e.g., Klein et al., 1975; Lloyd-Still et al., 1974). As was the case with respect to measures of cognitive functioning, it is essential that studies investigating the effects of malnutrition on social and emotional development attempt to develop the criterion measures on the basis of theoretical considerations. One way this can be done is to begin by asking the question, Is there prior research that suggests that certain behavioral functions are particularly vulnerable or sensitive to the particular nutritional variable being investigated? In developing the dependent measures for our research on the effects of early nutritional supplementation on children's social and emotional development at school age (Barrett et al., 1982), we considered both the research on the behavioral effects of protein-calorie malnutrition in animals and the recent research on the behavior characteristics of undernourished human infants. As will be shown in Chapter Three, both sources of information suggest that there is a distinct complex of behavior impairments associated with a history of undernutrition. Specifically, attention, state control, responsiveness to social stimulation, activity level, and tolerance for mild stress or frustration appeared to be affected. On the basis of these considerations, we formulated a set of hypotheses about the types of problems in social behavior and emotional development that one might expect to see at school age in children with a history of undernutrition. That is, we asked ourselves, If the kinds of behavior impairments identified above (i.e., in the animal and infant research) continued to characterize the child's early development, what would be the likely social and emotional consequences for the child by the time of early or middle childhood? On the basis of this formulation, we began to identify specific classes of behavior function relevant to nutritional history and then operationalized the variables. This research, summarized elsewhere (Barrett, 1984a; Barrett et al., 1982; Barrett & Yarrow, 1985; see also this volume, Chapter Four), has shown significant relationships between early caloric undernutrition and the child's social responsiveness, interest in the environment, capacity for active play with peers, persistence in a frustrating situation, and motor impulse control as measured in structured and nonstructured small-group situations at school age.

One of the most difficult and persistent problems in the research on malnutrition and behavior has been the problem of the confounding of constructs with levels of constructs, specifically with respect to the conceptualization and measurement of nutritional variables and the inferences drawn on the basis of these operations. For example, none of the major longitudinal intervention studies investigating the effects of nutritional supplementation on child behavior (Chavez & Martinez, 1979; Herrera et al., 1980; Klein, 1979; McKay et al., 1978) has been able to specify the range of supplementation over which effects of supplementation on specific behavior variables are likely or not likely to be identified. Further, there has been little attention paid to the question of linearity versus nonlinearity of supplementation effects with respect to specific behavior outcomes. There are several reasons why this particular problem of not differentiating construct from level of construct may occur. First, there are difficulties that arise solely from practical or ethical considerations, or both. In nutritional intervention studies, the investigators often are not in a position to manipulate levels of the independent variable simply because they are interested in identifying the parameters of specific nutrition-behavior relationships; rather, they are constrained by ethical considerations, social-political considerations, and the material resources available. In studies in which nutritional variables are not manipulated, it may be virtually impossible to obtain reliable data on the intake of specific nutrients. Under these circumstances, any plan to make inferences regarding effective levels of nutrient consumption or deprivation with respect to the behavior outcomes of interest may have to be relinquished and more general inferences about nutrition and behavior made. But it also is the case that a major cause of this problem is the failure of researchers to specify for themselves, to know beforehand, whether the levels of the independent variables that are being considered in their studies may be treated as randomly sampled levels from among a specified population of possible levels, or whether they represent fixed levels. The difference, of course, is critical with respect to the inferences that one may justifiably draw from the research. It seems reasonable to assume that researchers who are investigating the effects of malnutrition on human behavior by examining the influence of specific nutrient supplements on specific behavior outcomes do not wish to restrict their conclusions to inferences about the effects of one specific level, or perhaps a few discrete levels, of the treatment variable on behavior outcomes. Nonetheless, insofar as there is no attempt to address the issue of how levels of the independent variable were sampled from the possible levels, there is no justification for generalizing inferences about nutrition-behavior relationships to levels of the independent variable not included in the study.

A final concern with respect to the construct validity of our interpretations of empirical relationships between nutrition and behavior variables or the lack of such relationships is the problem of restricted generalizability across constructs. This is a potential problem both when a significant relationship between independent and dependent variables has been obtained and the investigator wishes to draw an inference about the theoretical meaning of the relationship and when there is no relationship identified and the investigator wishes to attach a theoretical interpretation to the lack of relationship. In the first instance, it may be the case that the validity of our inference that a particular independent variable has a specific effect on a particular outcome depends not only on a significant empirical relationship between the independent and dependent variables, but also on the assumption that other theoretically relevant potential outcome variables would have been similarly related to the independent variable, or perhaps even more strongly related, had they been included in the analyses, and that certain other variables would not have been related. For example, in a study of the effects of nutritional supplementation on children's cognitive functioning, the investigator might examine the relationship between nutritional supplementation and Wechsler Intelligence Scale for Children (WISC-R) IQ based on the hypothesis that malnutrition has a deleterious effect on attention capacities and that since the WISC-R includes subtests that appear to tap, to some degree, attentional capacity (e.g., Digit Span and Picture Completion), attention deficits should result in poor performance on the WISC-R and nutritional supplementation should result in improved WISC-R performance. If a significant relationship between nutritional supplementation and WISC-R IQ were obtained, the investigator might wish to interpret this outcome as indicating support for the interpretation that nutritional supplementation had a positive effect on attention and thereby improved WISC-R performance. The assumption that is being made, however, when one makes such an inference is that if a more precise measure of attention (i.e., one specifically developed to measure attention) had been included as a dependent measure in the analyses, an even stronger relationship between the nutrition measure and the behavior outcome would have been identified. If it were in fact the case that had such a measure been included, *no relationship* would have been identified, and that the significant relationship between the nutritional variable and WISC-R performance was due not to the effects of nutritional supplementation on attention but rather to the effects of nutrition on some other factor related to WISC-R performance, the investigator would be making a theoretical inference that was not justified by the data. When we consider the problem in this way, it is clear that restricted generalizability across constructs is a threat to any research study, since it is not possible to include in one's design and analysis all variables that might be expected to bear an empirical relationship to the independent variable under study (nor would it be desirable to try to do so, due to statistical considerations). It also is the case, however, that this threat to the construct validity of the interpretation of a study can be minimized if one includes in one's design and analysis one or more variables that on the basis of theoretical considerations could be expected to be related strongly to the independent variable, and that if found not to be so related would render questionable the validity of the investigator's interpretation of the relationship under study.

The converse of this situation occurs when an investigator finds no significant relationship between an independent and a dependent variable and concludes that no relationship exists, but fails to recognize that had other, theoretically relevant variables been considered in the data analysis, a different conclusion would have been drawn.

To illustrate this situation, let us again consider the situation in which an investigator is interested in examining the effects of a long-term nutritional intervention on children's growth and development. Let us also assume that the investigator begins with the assumption that before examining the effects of the intervention on behavioral development, it first is necessary to document that the supplementation has affected physical growth. The assumption is that if the supplementation is not significantly affecting growth, then it is not 'taking hold' and that, therefore, it makes no sense to spend time attempting to measure its effects on behavior variables.

This type of reasoning illustrates a common fallacy in nutrition research: That unless a nutritional manipulation—deprivation or supplementation—affects physical growth outcomes, it is not going to have an influence on other (e.g., behavioral) functions. For example, in a chapter on malnutrition and behavior, Ricciuti (1974) discussed the findings from the INCAP Guatemala study (p. 184):

> As the authors suggest, these preliminary findings may mean that the amount of additional protein taken in by these children was not sufficiently great to have any appreciable impact on

cognitive functioning as reflected in psychological test performance. To shed some further light on this question, it would be very useful to know whether variations in protein intake had any relationship to gains in height from age five to age seven. If the effects of dietary improvement were not sufficiently great to be discernible in improved physical growth during this particular age period, for example, one might wonder whether it is reasonable to expect significant improvement in psychological functioning.

While it may seem reasonable to assume in the case of a long-term intervention study such as the INCAP study that lasting effects on psychological function must necessarily be accompanied by significant effects on physical growth, it is not necessarily the case that a nutritional manipulation must influence physical growth before it can influence other functions and activities of the organism. Under certain conditions, manipulations of nutrient intake will change behavioral functions prior to influencing physical growth. In a condition of nutrient deprivation, body and tissue stores must be used to compensate for the deficiency in nutrient intake. At the stage of depletion of body tissue, the effects on the organism include not only the lowering of nutrient levels in the tissues (which can be measured by biochemistry) and changes in growth (measured by anthropometry) but also changes in enzyme levels and metabolic activities (Robson, 1978). The latter may influence the behavior of the organism including activity level, affect, and attention. Evidence for changes in nutrient levels influencing psychological processes prior to body size comes from studies of specific nutrient deficiencies, for example, iron deficiency, and studies of failure to thrive in infancy. Iron-deficient infants often are characterized behaviorally as showing little interest in their physical environments and being irritable. Positive changes in these affective and attentional characteristics occur within days after the institution of iron therapy, and this occurs before there is any significant change in hemoglobin (Oski & McMillan, 1981). Further, Oski and Honig (1978) found that when infants with iron deficiency anemia were treated with iron supplements, there were significant improvements in scores on the Bayley Scales of Motor Development Index, compared to control infants, as well as increased alertness, responsiveness, and sensorimotor functioning. Studies of failure-to-thrive infants, infants who have been nutritionally deprived in the context of inadequate caregiving and present with growth failure and clinical affective disturbances, have documented that improvements in social responsiveness as a result of nutritional intervention often precede changes in growth (see Rosenn et al., 1980). In sum, an investigator who fails to examine the behavioral outcomes of a nutritional intervention because the intervention does not appear to have influenced physical growth may miss the opportunity to identify significant behavioral effects of the intervention, with the result that his/her conclusion that the nutritional variable had no effect on the child will be invalid.

4.6 Conclusions and Recommendations

In this section I have examined potential methodological problems in conducting research on malnutrition and behavior. I have categorized problems using the classification scheme of Cook and Campbell (1979). Cook and Campbell identify four classes of threats to the validity of a research design:

statistical-conclusion validity,
internal validity,
external validity, and
construct validity of putative causes and effects.

In terms of research on the behavioral effects of malnutrition, these four types of validity may be defined as follows:

1. *Statistical-conclusion validity* is the extent to which the investigator can have confidence in the inference that an empirical relationship between specified measures of malnutrition and behavior does or does not exist.

2. *Internal validity* is the extent to which the investigator can have confidence in the inference that an identified empirical relationship between the measures of malnutrition and behavior represents, to at least some degree, a causal influence of malnutrition on behavior.

3. *External validity* is the extent to which the investigator can have confidence in the decision to generalize the nutrition-behavior relationship from the research sample to the target population; that is, the population of interest.

4. *Construct validity* of putative causes and effects is the extent to which the investigator can have confidence that (a) the theoretical interpretation of the nutrition-behavior relationship is consistent with the operationalization (measurement) of the variables, and (b) the operationalization of the variables is compatible with the underlying theoretical constructs and thus that the results of the study can be interpreted in terms of the theoretical framework in which the research issue was posed.

I have discussed and illustrated specific problems relevant to each of these four classes of validity and have given examples from the published research on malnutrition and behavior.

Unfortunately, the researcher investigating the effects of malnutrition on behavior is rarely in a position to control for the many threats to the validity of a study that I have explicated. First, unless a study is a true experiment—that is, involving random assignments to treatments under the control of the investigator, as are the major nutritional intervention studies—the majority of the threats to the internal validity of the study will be operative. A large majority of the published studies on the effects of malnutrition on behavior have involved retrospective or prospective comparisons of groups with natural histories of malnutrition versus adequate nutrition, and it is reasonable to assume that such studies will continue to be necessary to address certain issues (i.e., when it is necessary to determine whether a particular potential source of variance in behavior is significant under natural conditions). In such studies, the issue of internal validity must be addressed using the frequently inadequate methods of matching or statistical control (see Warren, 1973). Second, in most studies that do involve experimental manipulations, the investigators are not likely to be in a position from which they can control completely all the factors that might influence the utilization of the treatment, due to practical, ethical, and social-political constraints. For example, in the Bogota study (Herrera et al., 1980) discussed earlier, it was necessary for the investigators to provide supplements on a family basis without precise monitoring of individual intake. This resulted in differences among families in utilization of the supplement, which likely resulted in increased heterogeneity of response on dependent variables within treatment groups. In the Guatemala INCAP study (Klein, 1979), it was felt important to give individuals the choice of whether or not to attend the supplementation centers and with what frequency. This left open the possibility of the confounding of selection and treatment effects, or selection-treatment interactions, or both. Although INCAP investigators have not identified specific selection effects—specifically, maternal background or personality variables predictive of attendance at supplementation centers (Engle et al., 1979)—such effects cannot be ruled out. Third, the investigator studying the behavioral effects of a nutritional intervention may not be in a position, again for ethical reasons, to study the independent variable over a range of levels sufficient to address the important questions of (a) the amount of supplementation necessary to effect behavior change, and (b) whether there are nonlinear effects of supplementation on developmental outcomes. The researcher is likely to want to provide supplements beginning at a level of supplementation at which one can be confident that the supplement will have beneficial consequences for the subjects and to try to design the study so that nonfavorable effects of supplementation (as occurred, e.g., in the New York City supplementation project with the high-protein supplement group; see Rush et al., 1980) are avoided. Fourth, in many studies of the effects of malnutrition on behavior, researchers are interested in testing complex causal models involving large numbers of independent and dependent variables, not just one independent variable and one dependent variable. Not only are hypotheses frequently couched in terms of multiple predictors or multiple and correlated dependent variables (as in multivariate analysis of variance), but also a large number of significance tests generally are carried out. As a result there is, first, a seemingly overwhelming number of alternative causal relationships that have to be systematically considered (i.e., ruled out on a priori, theoretical grounds or empirically controlled or tested or both) and, second, a substantial increase in the probability of making incorrect probability statements and, specifically, type I errors. Finally, it is not always possible to anticipate, much less recognize, label, and conceptualize the many components of an intervention as it occurs in the field (i.e., components that were not part of the original theoretical conceptualization of the independent variable). For example, in a nutritional supplementation study that will involve personal interaction between investigators and subjects, it is difficult to know precisely what types of interaction will take place, how these interactions can be expected to influence the subjects, and, therefore, how one can best conceptualize these components of the treatment so that they can be built into the construct 'nutritional intervention.' Failure to account for these components will influence the construct validity of the investigator's interpretation of cause-effect relationships. Given the many potential problems of implementation, designing a study that meets even a majority of the threats to the statistical-conclusion, internal, external, and construct validity of a design would seem to represent a major achievement.

In the face of what would appear to be unattainable requirements for validity of research on the effects of malnutrition on behavior, investigators must attempt to devote their energies to those elements of the design the fulfillment of which is essential if the results of the study are to be theoretically significant. Thus, the position taken here is that although logically the hierarchy of threats to the validity of an experiment moves from statistical-conclusion validity (Was there a relationship?) to construct validity (How, exactly, may we interpret this relationship with respect to our hypotheses?), the priorities with respect to research design *must be* to identify one's theoretical issues, conceptually derive one's variables, and operationalize those variables in such a way that if and when an empirical relationship is identified, one can, with a high degree of confidence, interpret the relationship in terms of the theory or theoretical constructs that provided the basis for the research. In short, this means that utmost attention should be given to the many issues involving construct validity of putative causes and effects.

Adequate construct validity begins with identifying one's theoretical issues and then operationalizing both independent and dependent variables with a view toward (a) comprehensiveness and accuracy of definitions of independent and dependent variables, with respect to theory and prior research, and (b) careful operationalization so that operations correspond to conceptual definitions. It should be noted that in nutritional intervention studies, adequate conceptualization and operationalization requires giving special attention to the many 'hidden' aspects of an intervention (i.e., as it occurs in the real world) that might be anticipated and, it is to be hoped, built into the investigator's governing conceptual framework. It is also at this stage of design that the investigator must assess the theoretical importance of empirical issues involving levels of treatment, linearity versus nonlinearity of relationships, and related statistical concerns and decide whether such issues can be addressed in the study. Only when these essential considerations pertaining to the purpose and theoretical implications of the research have been thought through and resolved should the more strictly methodological issues of validity be addressed.

In making specific recommendations about priorities and requirements for research design, I first will suggest several guidelines that pertain solely to experimental nutritional intervention studies. This distinction requires further explanation, because the view is frequently expressed that experimental designs are inherently superior to nonexperimental designs (Tuddenham, 1974; Warren, 1973). Thus, one might expect that a first guideline would be to employ experimental designs whenever possible. As indicated earlier, this argument rests largely on the reduced threats to the internal validity of the study, resulting in greater confidence in causal inferences. I would suggest, in contrast, that the choice of a research design should depend primarily on the purpose of the study and that one should not assume that one type of approach is best suited for all purposes. For example, it is no doubt true that if the sole or primary purpose of the study is to demonstrate, as unequivocally as possible, a causal influence of nutrient intake on a particular developmental outcome, and in particular the possibility for enhancing behavioral development, an experimental design is essential. This is because, first, if the purpose is to unequivocally demonstrate a cause-effect relationship, experimental designs do reduce the number of potential threats to internal validity and, second, *logically*, experimentally manipulating nutrient intake provides the only conceptually valid, or at least potentially conceptually valid, means to address the research issue. Conversely, for certain purposes, naturalistic nonexperimental designs are preferable. For example, let us suppose that the primary purpose of the study is to examine two samples of retarded and nonretarded children to determine and compare the different odds ratios for risk of retardation due to such factors as early malnutrition, congenital disorder, parent psychopathology, and socioeconomic status. In such a case, a nonexperimental design would be more valid conceptually than an experimental design would be; one would not use an experimental manipulation of nutritional intake to address this research question. It is worth noting, in this regard, that under certain conditions, retrospective comparison analyses of data obtained from such a study are more powerful than prospective analyses of the data, a fact often overlooked by researchers in our field. This has been demonstrated empirically by Fleiss (1973). Consider the general case in which (a) associations are sought between a person's probability of being in a particular clinical, c, or nonclinical, $\bar{c}$, category on a presumed antecedent variable A, such as previous history of malnutrition or adequate nutrition, and a person's probability of being in a clinical or nonclinical category on a presumed outcome variable, B, such as mentally retarded or of normal intelligence; (b) the prior (unconditional) probability of clinical status on the outcome variable is lower than the prior probability of clinical status on the antecedent variable; that is, $p(B_c) < p(A_c)$; and (c) there is some true association between the variables; that is, $p(B_c/A_c) \neq P(B_c)$. Under these conditions, the probability of identifying, statistically, the association between the two variables for fixed sample size n is greater for a retrospective matched-comparison analysis than for a prospective matched-comparison analysis (i.e., given the same true relationship between A_c and B_c), and both approaches are more powerful than the naturalistic cross-sectional analysis (see Fleiss, 1973, p. 57).

To conclude this section, I suggest guidelines that the researcher should attempt to follow when designing a study to examine the behavioral effects of malnutrition or nutritional supplementation on human subjects. Each of these suggestions represents a conceptual or procedural activity on the part of the investigator to ultimately increase the construct validity, and thus the theoretical importance, of the research.

1. In examining the effects of a nutritional supplement or

treatment on behavior, the investigator should attempt to conceptualize the treatment variable with a view toward anticipating and building into the construct those nonnutritional components that necessarily will be included when the study is implemented. Thus, for example, theoretical inferences will be made about the effects of a particular nutritional, medical, social intervention, rather than simply a nutritional intervention.

2. In examining the effects of a nutritional supplement or treatment, the investigator will know how levels of the treatment variable were determined (i.e., do they represent fixed or randomly sampled levels?) and, in all cases, seek to include in the study those levels that permit the testing of the hypotheses of interest.

3. In examining the effects of a nutritional supplement or treatment on behavior, the investigator should, to the extent possible, administer treatments uniformly. If it is not possible to ensure that treatments are used in the same way by all subjects, data that would allow the investigator to test for differences in utilization and the effects of such differences on the relationships identified between independent and dependent variables should be collected.

4. It is essential that dependent variables be conceptualized and operationalized in a comprehensive and theoretically meaningful way. It is improper for an investigator to design a study without thinking through the questions of (a) What are the dependent variables of interest; what are the behaviors that, on theoretical grounds, one should expect to be related to the independent variable; and (b) What are the operations by which I can measure individual differences in these variables so that the measurements correspond to my theoretical constructs?

5. In testing a hypothesis about causal relationships between malnutrition and particular behavior outcomes, it is desirable to include in one's data analyses variables that although not the primary dependent variables of the study, can on theoretical grounds be presumed to bear significant relationships with the independent variable(s). If such variables are included in the analysis, failure to identify the postulated relationships provides cause for reevaluation of the original theoretical formulation.

6. To the extent possible, threats to the internal validity of the

research design should be controlled by employing techniques of randomization and control groups, which significantly reduce these threats. When, however, due to the nature of the study, such rigorous a priori control is not possible, and more imperfect statistical procedures must be employed, the investigator should reflect considerably on the likely contribution of such potential confounding influences to the results of the study and, to the extent that he/she believes that most or all alternative hypotheses can be handled within the theoretical framework in which he/she is working (i.e., ruled out a priori or tested and given meaningful interpretation within the theoretical framework), should proceed to design and carry out the research with consideration for the many conceptual issues that have been emphasized here.

7. In carrying out the research, attention should be given to the reliability of assessments, both nutritional and behavioral, with recognition that failure to do so may compromise the power of the data analyses and thus the investigator's ability to identify nutrition-behavior relationships. Particular attention should be given to assessing the reliability of behavior measures, since when the meaning of an assessment is different for the investigator and the subjects (as may happen when assessment procedures developed in one culture are used in another), it is unlikely that consistent and meaningful individual differences in responses will be obtained.

These recommendations, if followed, will increase the potential of a study for making useful contributions to theory and research on malnutrition and behavior. It is emphasized that more important than the specific type of design employed is the care with which the investigator has thought about the meaning of the variables to be investigated, the relationships that are hypothesized, and the broad theoretical network in which these relationships are thought to occur. When this careful conceptual groundwork is laid and the investigator takes the time and effort to translate the theoretically derived constructs into reliably measurable variables, the research that results is most likely to advance the scientific pursuit of knowledge regarding the role of malnutrition in behavior and development.

5.0 CATEGORIES OF RESEARCH DESIGN FOR STUDYING NUTRITION-BEHAVIOR RELATIONSHIPS (GENERAL CLASSES OF DESIGN)

There have been several different approaches used to study the effects of human malnutrition on behavior. The research, reviewed in Chapter Three, falls into seven categories: retrospective studies of severely malnourished children; malnutrition secondary to neonatal illness; effects of nutritional supplementation on children's behavior; studies of infants who are failing to thrive (NOFT); studies of intrauterine growth-retarded infants; correlational studies of chronically malnourished children; and clinical intervention studies.

In Chapter Three, in reviewing the literature for each type of study, we will describe the methodological features of each approach. In this chapter we consider the general classes of research design these specific approaches represent.

All the studies on malnutrition and behavior fall into one of the following categories:

1. Naturalistic, cross-sectional (correlational) designs;
2. Comparative, prospective designs;
3. Comparative, retrospective designs;
4. Experimental designs, involving random assignment to treatments;
5. Quasiexperimental designs, involving nonrandom assignment to treatments;
6. Longitudinal, sequential designs, involving multiple cohorts (experimental or quasiexperimental).

5.1 Naturalistic, Cross-sectional Designs

One of the most frequently used research designs in behavioral research is the naturalistic, cross-sectional design. Here, the investigator looks for the degree of relationship between variables of interest (e.g., nutritional history and later behavior) within a population, without selecting subjects on the basis of status on either variable. Generally, the degree of association is represented by a measure of correlation between the variables; that is, the extent to which individual differences on one variable are related to individual differences on the other(s). When qualitative data are collected, nonparametric measures of association are used. In studies on malnutrition and behavior, however, a strictly naturalistic approach generally is not used. This is because when there is a nonexperimental (i.e., nonintervention) design, the way in which one generally obtains information about the effects of malnutrition on later behavior is to compare groups with known and differing nutritional histories. This is particularly true for studies on severe malnutrition. There have been a few studies on chronic malnutrition, however, which have used a naturalistic approach; that is, examined how individual differences in behavior within a specified sample are related to individual differences in nutritional history or status. An examples is the study by Freeman et al. (1977) (reviewed in the following chapter) which related measures of anthropometry, presumably influenced by early nutrition, to behavioral functioning in children from an endemically malnourished population.

5.2 Comparative, Prospective Designs

It is important to distinguish between design for data collection and design for data analysis. For example, in many studies, data are collected on presumed independent and dependent variables at the same time, but the data are analyzed *prospectively*; that is, by grouping subjects on the basis of relative standing on the independent variable and then comparing them on the dependent variable. When this is done with data collected naturalistically, one has a prospective data analysis within a naturalistic design. For example, in a study by Cravioto et al. (1966) (described in Chapter Three), children from the sample were singled out if they were tall or short (highest and lowest quartiles on height) and then compared on behavioral measures. Although the analysis was prospective, the design still was essentially a naturalistic, cross-sectional design. Subjects were not selected for the study because of their heights; rather, groupings on height were post facto.

In a true comparative prospective design, subjects are chosen for the study because of their standing on the presumed antecedent (independent) variable and then compared on the dependent variable. Most of the studies to be reviewed on the effects of acute and early malnutrition on later behavior fall into this category. For example, the studies by Hertzig and colleagues in Jamaica (1972), Galler and colleagues in Barbados (1983), Stoch and colleagues in South Africa (1963, 1967, 1982), and Monckeberg and colleagues in Chile (1968) use this approach.

5.3 Comparative, Retrospective Designs

In biomedical research, a frequently used approach is to identify groups of persons, 'cases' who have a particular medical problem, and a comparison group, 'controls' who do not. The two groups are then compared for the likelihood that they had or did not have a particular experience earlier (presence or absence of 'factor'). For example, with regard to nutrition, one might compare children who were known to be developmentally delayed in some regard with a group who were not, and compare them for incidence of acute malnutrition in early infancy.

In general, the retrospective approach has not been used in the research on malnutrition and behavior. It has not been used in data collection (i.e., initially *selecting* children known to have a particular behavioral problem and comparing them with children without the problem with respect to early nutritional history), and it has been used infrequently in data analysis (within a cross-sectional or prospective design, choosing a subset of subjects with a particular problem and comparing them with a nonproblem group for incidence of previous malnutrition).[1] This may be unfortunate because, as shown earlier, the retrospective analysis design can be more powerful than the prospective analysis design under certain conditions.

[1] An example of a retrospective analysis appears in Klein, 1979, p. 235.

5.4 Experimental Designs Involving Random Assignment to Treatments

The classic experimental design involves selection of a group of subjects (with or without regard to prior standing on variables related to an outcome variable of interest), assigning them to treatments (values of an 'independent variable'), and comparing their performances on an outcome variable. In the present context, an experimental design would involve imposition of a nutritional intervention—either a systematic deprivation or a supplementation—and a study of consequent behavior. Virtually all of the animal studies, reviewed in Chapter Three, used the classic experimental design. In addition, a number of the nutritional supplementation studies for human mothers and infants used this design, for example, the studies by Rush et al. (1980) and Herrera et al. (1980).

5.5 Quasiexperimental Designs

In a quasiexperimental design, a treatment is imposed on experimental group members on a nonrandom basis. For example, treatments might be given to particular subjects who volunteer, and not to others. Another type of quasiexperimental design is one in which randomization is not by the individual but by some larger unit (e.g., a village or town). These larger units are assigned randomly to experimental and nonexperimental groups, but within units, subjects are not assigned at random to treatments. Thus, we have a quasiexperimental design (nonrandom assignment of subjects to treatments) with a randomized assignment of groups to treatments.

This design has been used in the nutritional intervention research. The most well-known example is the Guatemala study (Klein, 1979) in which villages were assigned to experimental (protein-calorie) supplement conditions or control (calorie only) supplement conditions. Although assignment of villages to treatments was random, individual assignment to treatments was not random. This study will be described in greater detail in the following chapter.

5.6 Longitudinal Sequential Designs, Using Multiple Cohorts

In the longitudinal sequential design, subjects are followed longitudinally, that is, over an extended period of time. The study design is invariably prospective; subjects are chosen for the study on the basis of prior standing on a presumed independent variable.

An elaboration of the longitudinal design (which may be experimental or quasiexperimental) is the multiple cohort, sequential design. Here, a number of different groups receive the same treatment, but the imposition of the treatment occurs at different time periods. For example, one group might begin the study and receive a treatment, X, at time T, while another group might receive the same treatment at time T+1. If the two groups were followed for three additional years (two additional years for the second group), the design would look like this:

	T	T+1	T+2	T+3
Group O	X	X	X	X
Group Y		X	X	X

The purpose of including a second cohort with the *same age* at treatment as the first would be to see if similar effects of the intervention occurred for children from different samples; that is, from the same population but (a) arriving at the criterion age at a different period in time; and (b) differing (presumably due to random factors) on other variables as well. It should be noted that in such a design, appropriate controls would be included. The Guatemala intervention study included several cohorts, that is, groups of mothers and children entering the study at different time periods, due to their becoming available at different times.

A related design also involves a sequential application of treatments but now on the basis of age rather than time period. That is, treatments are begun at different developmental levels, thus allowing the investigator to address the issue of the importance of the *timing* of the intervention. Again, the design may be either an experimental or a quasiexperimental design.

As an example, one might begin a nutritional supplementation at Age A for one group of children, and at Age A+1 for another. If such a design were continued for three years, the design would look like this:

	A	A+1	A+2	A+3
Group 1	X	X	X	X
Group 2		X	X	X

Both the Cali, Colombia study (McKay et al., 1978) and the Bogota study (Herrera et al., 1980) use variations on such designs. The major feature, then, of the design is that there are staggered or delayed implementations of the treatment. Thus we can see the differing effects of treatments begun at different times and/or sustained for different periods of time.

When the multiple cohort (different epochs) and multiple age level (sequential, age dependent) designs are combined, one has one of the strongest experimental designs (particularly if there is random assignment), in the sense that generalizability is increased. An illustration of a multiple cohort, multiple age level design for two age levels, two cohorts, and four years of treatment would be:

	T	T+1	T+2	T+3
Group O	A	A+1	A+2	A+3
1	X	X	X	X
Group O	A	A+1	A+2	A+3
2		X	X	X
Group Y	A−1	A	A+1	A+2
1		X	X	X
Group Y	A−1	A	A+1	A+2
2			X	X

Thus Group O1 would get four years of treatment, beginning at Age A. Group O2 would get three years of treatment, beginning at age A+1. Both groups would have been of the cohort turning age A at time T. The Y cohort would turn age A at time T+1. Group Y1 would get three treatments, beginning at time T+1 (age A). Group Y2 would get two treatments, beginning at time T+2 (age A+1). The analysis would, in addition to determining overall treatment effects (assuming inclusion of appropriate control groups), allow us to determine (a) whether the same effects obtain for a different sample (born at a different time) and (b) whether the timing and duration of the intervention is important.

Finally, it should be noted that it is possible to have multiple treatments; that is, multiple independent variables. For example, there may be an educational as well as a nutritional intervention. Both Herrera et al. and McKay et al. combined a multiple treatment (cognitive stimulation, nutritional stimulation, or both) into a multiple age level (delayed intervention) design.

6.0 SUMMARY AND IMPLICATIONS

In this chapter we have discussed four major concerns: (1) the importance of understanding the relationship between research design and theory; (2) the methodological issues in the measurement of behavior; (3) the threats to the construct validity of research designs; and (4) the categories of research design for research on malnutrition and behavior.

Our intent in this chapter has been to provide a basis for beginning to interpret the findings to be presented in the following chapter. First, we have attempted to convey the importance of asking, What is the theoretical issue that this study wishes to address, and is this study capable of addressing it? Frequently, a researcher purports to address a particular theoretical question concerning malnutrition and behavior, but the study, because of its design, does not make possible a test of that question. We have shown further how practical and ethical questions concerning the operationalizing of the variable "nutritional status" influence design and theory.

In the second section we have discussed different approaches to the assessment of behavior, in particular, observational methods. In discussing different approaches to data collection, we have shown that methods differ in: the extent to which they provide a basis for generalizability to different conditions of assessment; the degree of experimenter control over observations, and thus the probable replicability of assessments; their practicality; and the likelihood of eliciting individual differences in the behaviors of interest. In discussing differences in methods of data reduction, we have discussed the issue of the "level of analysis" and, specifically, the decisions faced by the investigator about the appropriate level at which to assign meaning to observed events. We also have addressed the question of whether procedures developed in one culture or population can be assumed to be usable and meaningful in another, and suggested guidelines for evaluating the cross-cultural validity of observational assessments.

The third section of this chapter is an attempt to describe in detail threats to the validity of research on malnutrition and behavior. Using the conceptual framework of Cook and Campbell (1979) we have defined and described the four major types of threats to the validity of a study: statistical conclusion validity, internal validity, external validity, and construct validity of putative causes and effects. For each general category we have described the specific types of threats that comprise each category and have given examples from the literature on malnutrition and behavior to show how these problems may occur. In making recommendations to the researcher, we have emphasized the primacy of the question of the construct validity of the research study: that is, is the researcher adequately addressing the theoretical question he or she is trying to answer?

Finally, we have defined and illustrated six major classes of research design: naturalistic cross-sectional designs, comparative prospective designs, comparative retrospective designs, experimental designs, quasiexperimental designs, and longitudinal sequential designs involving multiple cohorts. We have closed this section with these definitions and illustrations because in the following chapter, in which we review the literature on malnutrition and behavior, we will review the literature by type of study: for example, grouping retrospective studies of severely malnourished children, experimental supplementation studies, and cross-sectional studies of chronically undernourished children separately. In so doing, we will show the general class of research design that this type of study represents, and thus will be able to consider the strengths and weaknesses of the particular design in light of the information given here about the general class of research design it represents.

Chapter Three: Empirical Studies on Malnutrition and Behavior

1.0 OVERVIEW OF THE CHAPTER

Several sources of data converge to provide information about how a history of protein-energy malnutrition may influence the child's behavior and development. Much of the impetus for empirical study of the effects of malnutrition on children's behavior came from clinical observations of infants who were being treated for severe malnutrition. The consistent appearance of certain behavioral characteristics in infants with known calorie or protein-calorie deficits suggested the importance of more systematic, controlled study of children's behaviors. A second line of evidence has been the extensive work on experimentally induced malnutrition in animals. A large number of studies have shown long-term behavior impairments in animals systematically deprived of protein and calories early in life. The consistency of the syndromes associated with early malnutrition, across species and across methods of deprivation, raises the question of whether comparable behavioral effects can be identified in children. Finally, there has been, particularly in the last 20 years, research on the long-term behavioral effects of undernutrition in humans. Most of the early studies were nonexperimental comparison studies, examining behavioral differences between groups known to have experienced or not experienced severe undernutrition early in life. The majority of these early studies were concerned with effects of malnutrition on cognitive development, rather than on social and emotional well-being. More recently, there have been studies of the effects of nutritional supplementation on the behavior and development of children from endemically malnourished populations, that is, populations known to be at risk for developmental impairment due to undernutrition. By studying the effects of nutritional supplements—and, specifically, supplementation of the nutrient that is limiting in those populations—inferences can be made about the effects of chronic undernutrition on behavior. The supplementation studies have differed from the earlier studies, not only in design but also in that they have studied a broader range of outcome variables, including social as well as cognitive measures.

In this chapter we review the research studies that have led us to believe that chronic undernutrition may have long-term effects on children's behavioral development. The studies provide the empirical rationale for the study we will present in Chapter Four on the long-term effects of chronic malnutrition on children's social and emotional functioning at school age. Before beginning, however, it is important to note the following.

First, the focus here is on protein-energy malnutrition and not specific nutrient deficiencies. While recent studies have shown effects of deficiencies such as vitamin A and iron on children's behavior, the present review focuses only on protein-energy undernutrition.

Second, there are a number of methodological issues that raise interpretive questions about the results of most of the studies. In Chapter Two we discussed the major methodological problems and issues in research on malnutrition and behavior. In this chapter our focus is on the findings of the different studies. However, we also discuss methodological issues specific to certain categories of research study, and we discuss the implications of these issues for the interpretation of findings.

2.0 CLINICAL OBSERVATIONS

The clinical definitions of marasmus and kwashiorkor include the behavioral indicators of apathy, irritability, and sadness of expression (Robson, 1978). For example, Viteri (1981), in describing protein-energy malnutriton, writes, "Psychological alterations may be profound; severe anorexia, apathy, and irritability often make this child difficult to feed and manage" (p. 195). Srikantia (1982), describing kwashiorkor, writes, "The characteristic features of kwashiorkor are growth retardation, oedema and obvious mental changes which include apathy and irritability." And Gurney (1979) describes kwashiorkor and marasmus: The marasmic child "is commonly irritable or fretful but may appear apathetic.... The child with kwashiorkor is often miserable and apathetic, responding neither to affection nor simple stimuli, whether pleasurable or painful. The first smile is a good prognostic sign." (pp. 189–190).

These observations were noted in the early literature on severe protein-calorie deficiencies (Trowell et al., 1954; Gopalan & Ramalingaswami, 1955; Waterlow et al., 1960). Most researchers have interpreted these behavioral sequelae as necessary functional adaptations to the nutrient deficiencies. For example, Viteri views the affective and psychosocial characteristics of the severely undernourished child as adaptations which occur early in the disease process, and which delay, at least for a time, the loss of body tissue by decreasing activity. As pathology increases, with further deficits in lean body mass (muscle wasting), the reduced activity and affect become the consequences of the disruption in physical development. Gurney, too, sees changes in behavior, and particularly in activity, as primary adaptations to reduced energy availability. Later, however, they become part of a complex of physiological-behavioral adaptations which serve to maintain the child in a state of chronic undernutrition.

It is important to note that several of the early clinical reports on severely malnourished children suggested that the behavioral effects of undernutrition are closely linked to environmental deprivation.

Perhaps the most powerful description of the behavioral effects of undernutrition is that of Geber and Dean (1956) who studied kwashiorkor in Uganda. Geber and Dean examined 25 children who were unselected hospital admissions for kwashiorkor. Most were between one and three years of age. Children were given psychological tests from the Bayley scales, and their behavior was observed. In addition, their mother's behavior was observed. An important observation was that in the case of 13 of the 25 children,

ι separation of the child from its mother, consis-
ıstom and family tradition, with the child living

In testing, the children appeared extremely apathetic, but at times also highly resistent to the examiner, self-protective, and overly dependent on the parent. The authors said, "We realized the intense need of the child for the mother and his total dependence on her—a dependence so necessary that he had to refuse anything that might lead to a loss of contact" (p. 8). On a second testing, there often was improvement in children whose mothers had been attentive and caring. However, "...there was less evident progress in children whose mothers could not make them feel that they were safe" (p. 9). A third test, one month later, supported these observations.

The authors emphasized the link between separation and the child's depression following separation and the onset of kwashiorkor. "The consequences of psychological trauma (separation)," they noted, "are likely to be all the more severe the more the child is under-developed. The child who has been given security by maternal care and is, therefore, developing satisfactorily, may for a time be intensely distressed by separation, but the quality of the bond established with the mother will enable him to recover his balance as soon as he finds her again" (p. 12).

The interpretations by Geber and Dean are very important. First, because they identify so clearly the affective correlates of severe undernutriton. Second, because they emphasize the role of the caregiving environment in setting the conditions for or in preventing acute malnutrition. Third, because they suggest that the behavioral consequences of malnutrition—i.e., behavioral recovery—may depend largely on the response of the caregiving environment to the undernourished child. Each of these observations receives support in the empirical research.

It is important to note that more recent clinical studies of severely undernourished children arrived at similar conclusions about the role of the caregiving environment in maintaining malnutrition and/or moderating its effects.

Chase and Martin (1970) conducted a study of 19 children admitted to a Denver hospital for generalized undernutrition before the age of 1 year. The children were compared with a control group three to fours years later on measures of general development. The results of this analysis are summarized in a later section of this chapter. Relevant here are data on the mother's response to the question, "Did you want another child at the time this child was born?" More case mothers than control mothers responded "No," suggesting that maternal rejection may have played a role in the development of the illness. On the basis of this relationship and additional evidence that family stress and instability were related to acute malnutrition, the authors suggested the term "psychonutritional deprivation" to describe the condition of the undernourished children.

It is interesting to consider these observations in relation to those of Rajalakshmi (1975) on malnourished Indian children from well-to-do families. The author contrasts the adverse developmental outcomes of severe malnutrition in studies where the families were impoverished (Monckeberg, 1968; Stoch & Smythe, 1976) with the outcomes in three cases of severe malnutrition, where the children were from upper-class Indian families. All of the upper-class children were bottle-fed from birth or beginning at 2–3 months of age, and then received diluted baby food which apparently precipitated the malnutrition. All responded favorably to dietary treatment in the first year, and by school age (7 to 12 years) showed normal psychomotor development and school performance. Rajalakshmi noted, "In all three cases, the psychosocial environment of the home and affective care by the mother or foster-mother were more than satisfactory" (p. 361). The effects of malnutrition on behavior, he suggests, may depend not only on the type, duration, and severity of the deficiency, but also on the interaction of malnutrition and psychosocial factors.

We return to these observations and postulated relationships when we examine the controlled studies on malnutrition and behavior.

3.0 EFFECTS OF MALNUTRITION IN ANIMALS

Studies of malnutrition in animals have been important in helping scientists to understand the structural and functional consequences of malnutrition. In addition, they have enabled us to examine the persistence of adverse effects of malnutrition and the question of the environmental variables that may moderate those effects.

This section on animal research has three foci: First, effects of malnutrition on central nervous system development. This research has been particularly useful in helping us to generate hypotheses about critical periods for the effects of malnutrition on brain function. Second, effects of malnutrition on behavior, in particular, emotional responsiveness, exploration, and learning. These studies have demonstrated the range of behaviors vulnerable to malnutrition, and suggest that the malnourished organism suffers an impairment in the ability to attend to environmental cues and use information about the environment to make appropriate coping responses. A third focus is on studies examining the interaction of malnutrition and environmental deprivation (or stimulation) as they affect behavior and development.

Before proceeding with these topics, we examine the question of the appropriateness of using animal research to contribute to our understanding of the effects of malnutrition on humans. The question of whether one should "extrapolate" from findings from animal research to make inferences about malnutrition and human development has received much attention. Below we discuss the major issues.

3.1 Methodological Issues in Research on Malnutrition in Animals

In this section we will discuss some of the concerns frequently raised about the methodology in animal research and, in particular, the justifiability of using animal studies to further our understanding of the behavioral effects of undernutrition in humans. These concerns are:

1. The question of the comparability of similar behaviors in different species;
2. Species differences in central nervous system development;
3. Comparability of methods for producing malnutrition and, in particular, the confounding of malnutrition with stimulus deprivation; and

4. Confounding of long-term effects of severe malnutrition in early life with the concurrent effects of malnutrition.

3.1.1 The problem of the meaning of behavior

Morrison and McKinney (1976) argue that "similarity of behaviors in different species does not imply similarity of mechanisms or meanings of those behaviors." (p. 134) They also question whether the concepts of normality and abnormality have the same meaning in animal and human behavior. Two responses must be made to these criticisms. First, it is true, not only for human-animal comparisons but for within-human comparisons, that the same behaviors do not always have the same "meanings" (i.e., functions and mechanisms) even when exhibited by members of the same population. When there is evidence of the same associations between a particular behavior and other conditions—e.g., certain stimulus conditions or correlated behaviors—in different populations, we may have more confidence that the behavior has the same "meaning" in the different populations. This is a test of construct equivalence: If a behavior has the same correlates in one population as it does in another, we have more confidence in treating those behaviors as equivalent than we would if the correlates were entirely different (see Barrett, 1982).

The second point is that if we take a probabilistic (i.e., empirical) approach to the definition of abnormal behavior, we can speak of abnormality in the same way for animals as we do for humans. It seems reasonable to define an abnormal behavior as a behavior that (a) has a low probability of occurrence under normal conditions of growth and development, and (b) is associated with behavior patterns and/or responses from the environment that prevent the organism from being integrated into the social community.

3.1.2 Species differences in central nervous system development

A major concern about "extrapolating" from research on animal malnutrition in order to generate hypotheses about the effects of human malnutrition is the fact that there are species differences in central nervous system development (Morrison & McKinney,

1976). In particular, there are differences across species in the periods of rapid cell division and cell enlargement in the brain and other organs. (See the following section on organic impairment resulting from malnutrition.)

It is now believed, however, that the patterns of central nervous system development are far more similar across species than they are different. Dobbing, who introduced the critical period hypothesis regarding the effects of central nervous system insult resulting from malnutrition summarized the matter as follows:

"...the development of the human brain is essentially so similar to that of animals that it is reasonable to expect help from the study of rats in elucidating the human condition. The basic stages of brain development are virtually identical in all mammals.... All the species go through the same developmental sequence, and the anatomical units of which the brain is composed are also remarkably similar in chemical composition and in metabolic and electrophysiological properties from one species to another." (p. 98, 1974)

Further, after describing at length the major differences between species, which is in the timing of the growth spurt for cell division, and its implications for cross-species research, Dobbing concluded, "...animal experiments, properly conducted and responsibly extrapolated, do illuminate investigation of these matters" (i.e., the effects of malnutrition on human development), although "we must all be much more careful about our interspecies extrapolations" (p. 99).

3.1.3 Procedures for producing malnutrition

Animal studies on the effects of malnutrition have used several methods for producing malnutrition, including limited daily nursing, increased litter size, postnatal protein deprivation, and maternal protein deprivation. Such methods have no parallels in human studies, which have either employed nonexperimental comparison designs or experimental supplementation designs.

Furthermore, animal studies that involve maternal malnutrition or preweaning malnutrition often involve maternal deprivation; that is, a reduction in the sensory and social stimulation normally afforded the infant by the parent. Levine and Wiener, in their critical review of the research on malnutrition and animal learning (Levine & Wiener, 1976a), emphasized the confounding of nutritional and extranutritional variables in the experimental animal research. Smart (1984) suggested that if the mother-infant relationship is impaired by the experimental nutritional manipulation, the interactive disturbance itself may jeopardize the young animal's development.

These issues may be addressed. First, it is the business of science to identify relations between variables and, more often than not, the conditions for assessment are not identical. Experimental methodology makes possible greater control over situational factors, but scientists must still attempt to draw inferences about relationships by using information obtained in different ways and from different sources. Thus, it is not a problem specific to nutrition research that generalizations about relationships between events must be made despite differences in the measurement conditions of the events, and despite differences in the surrounding circumstances. Second, it is the very fact of having to "transcend" interexperiment, interindividual, and intersituational differences that makes possible the validation of our interpretations. If it is the case that "A" relates to "B" under a variety of experimental conditions—that is, if the relation is robust over a variety of experimental conditions—we may have more confidence in our interpretations of the meaning of the relation.

With regard to the concern about mother-infant interactive disturbance as a "confounding influence", there are two responses. First, malnutrition must influence behavior through some mediating processes. Whether one conceptualizes the processes as purely physiological or as mediated by the effects of the change in nutrition on certain interactive behaviors between mother and infant (these effects themselves due, ultimately, to a physical insult), the effect is still interpretable as relating to the nutritional variable. Second, as Fleisher and Turkewitz (1984) have noted, in human infants as in many animals the infant's nutritional status affects the mother's response to the child. While in humans the factors influencing maternal behavior are more complex than for animals (e.g., social environmental variables interact with child and parent biological variables), it is likely that the developmental effects of early malnutrition are mediated largely by changes in mother-infant interaction (see Galler et al., 1984, for a review). In this sense, experimental models of animal malnutrition provide models for the effects of malnutrition on behavior that are not unrelated to the postulated processes whereby human malnutrition may affect the child's later development.

It is interesting to note that there is now evidence that either experimentally imposed maternal malnutrition (Crnic, 1980) or malnutrition imposed on animal infants (Fleisher & Turkewitz, 1981) may result in animal mothers showing more intense maternal behavior—for example, retrieving, nest building, and nursing—than they would otherwise. These effects may, under some conditions, compensate for the effects of the malnutrition on the animal young and thus lead us to *underestimate* the direct effects of the reduced nutrition on the developing organism. Clinical studies of failure to thrive infants have similarly documented cases of overly stimulating and nurturant behavior of mothers toward these infants.

3.1.4 Long-term vs. concurrent effects

The fourth problem is the confounding of long-term effects of malnutrition (i.e., for later development) with the effects of on-going nutritional deficit. In most studies, the researchers have not provided the controls necessary to address the question of whether a particular behavioral effect was due to (a)the lasting effects of early undernutrition, (b)impaired performance due to current undernutrition, or (c) both a and b.

The issue is an important one because a major concern in many human studies of early and severe undernutrition is that the previously undernourished are also chronically and currently undernourished (see Frank, 1984). We discuss this problem with reference to human malnutrition in Chapter Five.

3.2 Effects on Central Nervous System Development

The critical period hypothesis of central nervous system development implicitly or explicity influenced the design of much of the research on animal and human malnutrition summarized in subsequent sections of this chapter. This hypothesis suggests that there is a chronologically fixed period when the brain, or any organ, possesses transient biosynthetic abilities to generate new tissue, abilities that will not be present in later life. According to this hypothesis, if nutritional substrate is not available at the time required by the developmental schedule of organogenesis, deficits and distortions in brain structure will evolve that will not be remediated by provision of adequate substrate in later life, when the biosynthetic processes to generate new tissue are no longer active (Dobbing, 1985). The corollary of this hypothesis is that the degree of nutritional deprivation necessary to produce adverse consequences is inversely proportional to the rate of development at the time of the insult (Dickerson, 1981). The effects of malnutrition occurring after the period of most rapid brain growth would be reversible, by nutritional resuscitation, while the effects of early malnutrition would not.

The critical period hypothesis was further elaborated to suggest that malnutrition during the period of hyperplasia, when the cell number was increasing, would result in permanent deficit in cell number, and thus permanent deficit in organ size (Winick & Noble, 1966). It was assumed that increase in cell size, or hypertrophy, only occurred after increase in cell number. Deprivation during the period of hypertrophy was thought to create potentially reversible deficits in cell size (Winick, 1976). Thus the most vulnerable period of brain growth was thought to be the period of most rapid cell division.

The period of most rapid brain growth in the central nervous system varies from species to species. For example, in the guinea pig, it is prenatal; in the rat, during the period of suckling; and in the human from the third prenatal trimester through the second postnatal year (Dobbing, 1973). Thus, in seeking to elucidate the effects of nutritional deprivation on human brain development, animal researchers have in general timed the deprivation to coincide with the expected critical period of the species under study, and the onset of rehabilitation with the expected end of the critical period.

The critical period hypothesis is supported in broad outlines by extensive neuropathological evidence in both animals and humans, for which the reader is referred to a number of reviews: Dickerson, 1981; Winick and Morgan, 1985. At the grossest levels in human children, early malnutrition results in deficits in brain weight in children dying from undernutrition (Winick et al., 1972; Winick, 1976) and in deficits in head circumference in those surviving (Hoorweg, 1976; Stoch & Smythe, 1976). The assumption that increase in cell number precedes increase in cell size, and that functional deficits inflicted are most serious only if cell number is affected, has been modified by recent research. As Dobbing (1985) has emphasized, the development of the brain is heterogeneous. The brain does not in a uniform fashion complete hyperplasia and then begin hypertrophy. Rather, cell replication in one sector may occur concurrently with cell enlargement in another. Moreover, the processes in the brain vulnerable to nutritional deprivation have been found to extend beyond simple cell replication and cell enlargement to include myelination, and the maturation of myelin content, the formation of synapses, and the degree of dendritic arborization (the formation of interconnections between neurons) (Dickerson, 1981; Winick and Morgan, 1985). These processes continue beyond the boundaries of the initially specified critical period and may, as Dobbing (1985) has pointed out, have greater functional significance than simple reduction in cell number. Myelination, for example, facilitates the transmission of nerve impulses; the extent of dendritic arborization affects the capacity for integrated dissemination of stimuli throughout the brain (Dickerson, 1981). Additionally, it can no longer be assumed that malnutrition occurring after brain structure is established will have no ill effects on brain function. As discussed in Chapter 5, because malnutrition reduces brain levels of amino acids, which are the precursors of neurotransmitters, brain function remains vulnerable to nutritional insult after the growth spurt is completed (Dickerson, 1981).

While the neuropathological sequelae of early malnutrition are more complex than simple deficits in brain size, there remains considerable uncertainty as to the behavioral consequences of structural deviations. Particularly in the immature brain, there is as yet no precise correlation of observed behavior with the myriad anatomic and physiological developmental changes. Moreover, since anatomic data in humans is of necessity from nonsurvivors, it is not clear whether processes such as synaptogenesis and dendritic arborization can, unlike cell number, recover from early nutritional insults if appropriate nutritional rehabilitation is provided (Dobbing, 1985). It also is not clear what the functional significance of such recovery would be.

Finally, it is important to emphasize that the identified structural consequences of early malnutrition do not suggest that deficiency of nutritional substrate alone explains the observed structural deficits following early undernutrition. As is discussed later in this chapter and in Chapter 5, sensory stimulation may also influence the development of central nervous system size and structure (Greenough, 1975). Even when considering neuroanatomy, malnutrition cannot be considered a "pure" insult, but must be evaluated in the context of the environment in which it occurs, and the malnourished organism's interaction with that environment.

Much of the animal and human research summarized below was designed to evaluate the effects of nutritional deprivation during the implicitly assumed period of maximum central nervous system cell multiplication specific for each species, with little expectation that significant detrimental and rehabilitative effects of nutritional experience might extend beyond this period, or that non-nutritional factors might also affect the central nervous system. Some of the apparent inconsistencies in the early research may be explained by the relatively simplistic model of brain development that guided its design, a model that no longer reflects current thinking in the field. A more current interpretative model will be discussed in Chapter 5.

3.3 Behavioral Effects of Malnutrition in Animals

Particularly relevant to the research on malnutrition and behavior in humans are the studies that have investigated the effects of early malnutrition on learning, emotionality, and social behavior in primates and subprimates. In general, such studies have indicated reductions in exploratory behavior, attentiveness, and persistence in problem-solving tasks, and heightened emotionality, fearfulness about new situations, and irritability following nutritional deprivation.

In most reviews of research on malnutrition and animal behavior, studies are considered together under the general rubric "behavioral development". This practice obscures important differences in the functions being explored. Further, it makes it difficult to identify consistencies and inconsistencies in the findings. In the present review we discuss the effects of malnutrition on animal exploration, emotionality, social behavior, and learning, respectively. We then discuss consistencies and inconsistencies in the research findings. Finally, we examine the role of environmental stimulation or deprivation in moderating the effects of malnutrition on later behavioral development.

3.3.1 Exploration

The data on the effects of malnutrition on later behavior in animals are very consistent in one area: Animals malnourished prenatally and/or postnatally show reduced exploration of a novel environment at the time of malnourishment, and the effects on exploration tend to persist.

The functional importance of exploration is that it indicates the animal's interest in learning about the environment, and it has implications for the animal's ability to adapt to new situations. Exploration reflects the animal's capacity for attention to stimuli and stimulus change, and thus any insult that compromises the animal's capacity to attend to the stimuli around it will influence exploration.

In order to study exploration, a situation should be used that is novel and interesting to the healthy animal, but not so novel and provocative that it is frightening. If the animal is placed in an extremely strange situation, for example the open-field situation, the resulting emotional response—which may be one of heightened nonpurposive activity or drastically decreased activity—will make it difficult to interpret the animal's behavior as indicating high or low exploration.

Several approaches have been used. For example, Levitsky and colleagues (see Levitsky, 1979) observed malnourished and well-nourished rats in a playroom with toys, and examined the number of contacts with both other animals and toys. Zimmerman and colleagues (see Zimmerman et al., 1974), studying the behavior of malnourished monkeys, have examined exploration by watching monkeys make contact with toys and other monkeys in a playroom. In addition, they have looked at preference for viewing complex stimuli by giving animals an opportunity to put their heads through a partition to view slides of different degrees of complexity. A similar approach was used by Wiener and colleagues (Wiener, Robinson, & Levine, 1983) as they studied the extent to which rats would poke their noses through a hole in the floor to explore. Zimmerman and colleagues also have examined exploration by studying the animals' interest in pulling a chain to reach a novel object.

On the basis of observations of exploratory behavior in rats (Levitsky 1979) and pigs (Barnes et al., 1976), Levitsky (1979) concluded that animals severely protein-calorie restricted postnatally show a significant decrease in their "hunger to learn." The reduced interest in the environment has the effect of reducing the animal's capacity to learn about the environment in situations in which the animal has strong competing responses. As we discuss in the subsequent section on learning (see Sec. 3.3.4), this deficit in the motivation to learn results in poorer performance on tasks in which attention to incidental stimuli is important for learning.

Zimmerman, Geist and Wise (1974) summarized a series of programmatic studies on the effects of postnatal malnutrition. Reductions in exploratory behavior and/or approach behavior occurred in the first year, concurrent with protein-calorie malnutrition (Zimmerman & Strobel, 1969; Zimmerman et al., 1972, 1974); in the first and second year also during the period of malnourishment (Strobel, 1972; Strobel & Zimmerman, 1971); and after two years of age (Aakre et al., 1973; Strobel & Zimmerman, 1972; Zimmerman et al., 1970).

We discuss one of these studies in detail to highlight the differences in exploration between adequately nourished and malnourished animals.

Zimmerman et al. (1970) examined neophobic behaviors in malnourished monkeys—behaviors indicating phobic avoidance of novel stimuli. A common observation of normal monkeys is that they tend to approach new objects placed within their reach when the situation is familiar (Butler, 1965). The investigators examined monkeys who had been separated from their mothers shortly after birth and placed on a 22 percent protein diet. Later in infancy, the diet was reduced further in its protein. Animals were housed in individual cages. After approximately 11 months on the protein-deficient diet, animals were exposed to a problem-solving situation in which the correct response called for an object discrimination-reaching response. On the last of a series of trials, a novel object replaced the familiar object (either for the correct or incorrect stimulus), and the increment or decrement in performance was examined.

The groups did not differ significantly on overall or trial-by-trial performance through the first six trials. In fact, the low-protein monkeys had consistently higher performance scores on each trial. On the seventh trial, however, the one in which the novel stimulus was introduced, the animals showed a dramatic change in behavior. The malnourished animals made significantly fewer approach responses to the novel object. This difference was not a function of accuracy of discrimination. The high-protein monkeys made twice as many responses as the low-protein monkeys when the novel object replaced the positive stimulus in the discrimination problem, and over three times as many responses when the novel object replaced the negative stimulus. The most parsimonious interpretation is that the low-protein monkeys showed greater inhibition in the face of a novel stimulus, reducing normal exploration.

Other animal studies showing reduced exploration at the time of the malnutrition were Simonson and colleagues' studies (Simonson, 1979) of passive avoidance learning (cited in the Learning section, below), and Frankova's studies on social behavior and the interaction of malnutrition and stimulation (1974) also reported in a later section.

A recent study showing effects of malnutrition on exploration is Wiener and colleagues' study (1983), referred to previously and to which we will refer again when we discuss malnutrition and emotion, and malnutrition and learning. In this study, Wiener et al. found that there were lasting effects of prenatal and early postnatal malnutrition on exploration. Malnourishment was achieved by having pregnant rats placed on a low-protein diet and maintained on the diet through lactation. When the infant rats, nutritionally rehabilitated after weaning, had grown to adulthood, they were tested in a number of situations, including one in which they could explore a novel environment by poking their noses through a hole in the floor. While malnourished and well-nourished rats did not differ on a number of other measures—including activity in the open field—there were significant differences in exploration, with previously malnourished rats showing decreased exploration.

3.3.2 Emotionality

Studies on malnutrition and emotionality in animals have been very important in generating hypotheses about the behavioral effects of malnutrition in humans. The early clinical case studies of malnourished children identified affective disturbances in severely malnourished infants and children. The animal research has generally identified effects of malnutrition on animal behaviors which could be described as reflecting emotion or state of arousal. This research is reviewed below.

It should be noted that significant effects of malnutrition on attention, emotion, or state of arousal have important implications for the organism's functioning. Performance on simple learning tasks and the ability to learn complex tasks depend on attentional and emotional factors. Disturbances in emotional functioning would have implications for behavior in a wide range of areas.

The primary method used to study emotional reactivity in animals is the open-field test. In the open-field test, the animal is placed in an open field—an area that may take any of a number of dimensions—and the animal's motoric and autonomic activity are studied. Activity level, consummatory behavior, and frequency of defecation are the most frequently used measures of emotionality. Denneberg (1969) has shown how the open-field test and the measures of activity level and defecation have construct validity as measures of emotionality.

A number of studies have shown that severe protein-calorie restriction in animals results in heightened emotionality. Cowley and Griesel (1964) showed that first filial- generation rats of malnourished mothers who were weaned onto low-protein diets were more emotional than control rats during the period of malnourishment. Simonson et al. (1971) also found greater emotionality in malnourished rats than control rats. Animals in the study were prenatally malnourished and were compared to control rats at 26 weeks. Prenatally malnourished rats showed less activity and more defecation than control rats and, in a follow-up at 76 days, the differences were maintained. In addition, animals who also had been postnatally deprived were the most emotional.

Levine and Wiener (1976b) have pointed out that there are inconsistencies in the literature on malnutrition and emotional response. They point, for example, to the study by Smart (1974) showing that prenatal malnutrition resulted in increased activity and reduced defecation, results opposite to prediction. In the Smart study, however, frequency of rearing was used as a measure of activity. In other studies, rearing or vertical activity has been interpreted as showing emotionality (Frankova & Barnes, 1968; Guthrie, 1968).

Studies that have looked at other measures of emotion have shown clear relationships between pre- and/or postnatal undernutrition and emotional response. These studies have examined the animal's ability to cope with a frustrating or stressful situation, and have looked at, as outcomes, excitability, activity, and nonpurposive behavior.

Barnes et al. (1968) restricted rats' protein and calorie intake postnatally, through the period of lactation following weaning. Feeding was restricted to one hour per day. In contrast to animals without a history of malnutrition, the previously malnourished rats showed high emotionality: They spilled more food and were more excitable than the other animals.

The data of Frankova and Barnes (1968) on active avoidance learning also are relevant. Rats malnourished prior to weaning and following weaning performed as well on an active avoidance task as did nondeprived rats. During the intertrial intervals, however, while waiting for the conditioned stimulus that indicates the imminent onset of shock, they showed heightened emotionality and excitability, manifested by abnormal jumping behavior.

Two recent studies help contribute to our understanding of the nature of the link between malnutrition and emotional response.

Gallo (1981) examined the consequences of prenatal maternal protein deprivation and protein deprivation during lactation on the open-field activity of rats. Animals were tested in the open field just prior to puberty. Restricted animals showed lower activity levels than controls and initially lower levels of rearing behavior.

Associated with the lower activity was reduced basal and stress levels of corticosterone. The results indicated a suppression of responsiveness to a normally stressful situation.

Wiener, Robinson, and Levine (1983) examined the behavioral and pituitary-adrenal response to novelty in malnourished rats. Rats were prenatally and postnatally malnourished by having pregnant mothers placed on low-protein diets, on which they were maintained through lactation. Control rats were not malnourished.

Animals were studied in the open field and in an exploration situation. Results of these studies were reported above. In addition, behavioral and pituitary responses to increasing levels of environmental novelty were studied. Both malnourished and well-nourished animals showed increases in plasma corticosterone levels with increased changes in their environments (i.e., in relation to their home cages). A major difference, however, was that previously malnourished animals were not able, as were well-nourished animals, to use a consummatory activity—drinking—to modulate and reduce the endocrine activity. The authors interpreted this outcome as showing a diminished capacity to "effectively utilize environmental stimuli to make appropriate coping responses." (p. 49) We return to this study later in our discussion of the consistencies among findings on malnutrition and animal behavior.

3.3.3 Social behavior

Data on the social behavior of malnourished animals also indicate heightened emotionality. Studies of dogs, monkeys, and rats link severe protein-calorie restriction with heightened aggression, negativity in response to other animals' initiations, and generally unpredictable behavior.

Platt and Stewart (1968) administered diets of low and high protein value following weaning to dogs born and suckled of well-nourished mothers and to dogs who were congenitally malnourished. Animals born of malnourished mothers and given diets of low protein value were unpredictable and apprehensive in their behavior as adults. They did not engage in normal play activities and often were aggressive toward humans and other animals. Behavioral abnormalities also were noted for animals born of well-nourished mothers, but later subjected to protein restriction.

Zimmerman et al. (1972) examined the social behaviors of monkeys first subjected to low-protein diets at 7 months of age. When they were approximately 10 to 20 months old, the low-protein monkeys showed less social-tactual and social-approach behavior than the high-protein animals, and spent a greater proportion of their time engaged with nonsocial objects or in nondirected behavior. Low-protein monkeys also were more aggressive than control monkeys, aggression being measured relative to overall social responsiveness. The authors noted that malnourished monkeys consistently failed to make normal responses to the social initiations of other animals.

Frankova (1973b) examined the social interactions of rats who had been subjected to preweaning and/or postweaning protein-calorie restriction at ages 9 to 27 days (i.e., prior to weaning) and at 38 days, the end of the postweaning deprivation period. Results obtained during the preweaning period indicated that beginning at day 9, protein-restricted animals made fewer active contacts with littermates and were more often in isolated activity than controls. When behavior in the rearing cage was examined, beginning at day 15, protein-restricted animals showed less play with littermates, less active aggregation, and less play with the mother than the well-nourished rats. Beginning at day 16, social behavior was examined by pairing the rats in a new environment. Malnourished rats were more aggressive than well-nourished rats, engaged in less grooming, and were more negative in response to approaches from littermates. Social behavior observations following weaning indicated a tendency for protein-deprived rats to respond to the presence of another animal with withdrawal and inhibition of exploration.

A recent study by Gallo and colleagues suggests an effect of malnutrition on the social behavior of the animal toward its mother. Gallo, Werboff, and Knox (1980) examined the behaviors of prenatally and postnatally malnourished kittens toward their mothers in an "attachment situation". Kittens whose mothers were protein restricted during pregnancy were then undernourished postnatally, by having the mothers protein restricted during lactation. These cats were compared with nonrestricted cats at 2, 9, and 10 days postparturition in a situation in which they were first placed with the mother in a holding box, and the mother was then removed. Protein-restricted cats showed more crying than controls, regardless of the mother's proximity, and less activity. Disruptions in maternal attachment behavior, specifically, retrieving and comforting, also occurred in the experimental group.

In summary, the studies on exploration, emotionality, and social behavior show that malnutrition has an effect on the animal's capacity to tolerate stressful stimulation and to make appropriate coping responses. Severely protein-restricted animals are less interested in their physical environments, less capable of appropriate social response, and less successful in adapting to changes in their environments and making constructive responses to stressful problem-solving situations than well-nourished animals. These outcomes are very important in suggesting the areas of behavior function that may be affected by malnutrition in humans. We show, in Chapter 4, how these findings on animal malnutrition have been important in generating hypotheses about the long-term effects of human malnutrition.

3.3.4 Learning

No area of research on malnutrition and behavior is so difficult to summarize and synthesize as the research on malnutrition and learning. There is a great deal of disagreement about the long-term effects of malnutrition on the capacity to learn, and researchers frequently express frustration at the apparent inconsistency of findings (cf. Levitsky, 1979).

Some of the methodological issues that may contribute to the inconsistency have been noted, including, in particular, differences in the methods for producing malnutrition.

The problems are even more complex, however. The problems in relating malnutrition to learning primarily revolve around the problem of separating—i.e., conceptually and methodologically—behaviors reflecting "capacity" and behaviors reflecting "performance" or "motivation." Obviously, learning—an observable change in behavior related to experience—will reflect the animal's motivation to learn. Any insult that compromises attention and motivation may affect learning. But an insult or experience that decreases attention and motivation to learn under some circumstances may increase it under others. Experiences that cause a narrowing of attention under some circumstances may result in a sharpening under others. And behaviors that appear to reflect poor attention to certain stimuli may simultaneously reflect high attention to others.

As an example of the problems in making predictions about the effects of experiences that influence attention and performance on the capacity to learn, consider what happens when children are challenged to the degree that they become anxious about successful performance. Given a routine task and a high degree of pressure to succeed, children may show increased efficiency of performance on tasks requiring the performance of "overlearned" behaviors such as copying, placing pieces in holes, or coloring in forms. Because of the effects of anxiety on attention, however, children may have a hard time with less routine, more challenging activities requiring creative thinking and novel combinations of behaviors. Thus, it is a truism that under stress children will perform well on routine tasks but not well on very difficult ones. (Think about what kinds of work we choose to do when we are feeling upset or "stressed" versus when we are feeling relaxed and happy.) The same experience that may lead to failure to learn a difficult task may lead to improved performance on a simple learning task.

This, of course, is only one way in which the relation between motivation and learning poses theoretical and interpretive difficulties for the researcher studying malnutrition and learning capacity. For example, another problem that will be elaborated in the following discussion concerns the confounding of food deprivation (and the resulting physiological changes) and hunger (i.e., the experience of wanting food) in interpreting the effects of food deprivation on behavior.

Despite these problems, we see a thread that runs through the literature on malnutrition and learning, and it is closely related to Wiener et al.'s observation about malnutrition and impaired "coping." A number of studies—in particular, recent studies—suggest the following: First, early malnutrition results in performance deficits on tasks in which the animal must attend to incidental stimuli in the environment and/or must inhibit emotional responses to stressful situations. Second, negative effects on performance and learning may not be observed if the animal must perform a response that is highly intrinsically motivated (e.g., a response related to survival) or if the animal has been motivated to perform by recent underfeeding. Third, there may be long-term effects of undernutrition on learning capacity. These would reflect the cumulative effects of learned behavioral adaptations to poor diet.

In the following sections we review specific studies on malnutrition and learning, referring to the methodological and interpretive issues raised above. We conclude with a restatement of the major conclusions concerning both areas of consistency and areas of inconsistency in the findings.

3.3.4.1 Studies on learning A number of studies have shown that severe malnutrition in animals results in poorer learning on tests of visual discrimination, spatial relationships, and avoidance conditioning.

Barnes et al. (1966) examined the effects of pre- and postweaning malnutrition on the maze learning of rats. Two groups were protein restricted prior to weaning, and two groups nonrestricted. Within each group, one group was assigned to a low-protein diet postweaning, from three to eight weeks of life. In general, postweaning deprivation resulted in poorer performance on the water maze task, and the doubly deprived rats performed at the lowest level. Rats subjected to nutritional deprivation at either phase performed more poorly than controls on both tasks.

Levitsky and Barnes (1975) examined the maze learning performance of rats fed low protein diets after weaning. In absence of a food reward, malnourished animals showed less active exploration of the maze than well-nourished animals. Having failed to learn the maze under no-food conditions, the experimental animals showed no improvement in performance when retested in the same maze under food reward conditions.

Strobel and Zimmerman (1971) examined the problem-solving capacities of monkeys subjected to low-protein diets at 7 months of age and tested at 14 months of age. Protein-deprived monkeys were able to learn complex three-part puzzles as readily as control monkeys. On a simpler, two-part puzzle administered at 8 months of age, low-protein monkeys showed a loss of interest in the problem, manifested by rapid improvement across the first two trials, and a failure to improve thereafter.

In an attempt to separate motivational and attentional factors in performance from cognitive abilities, Zimmerman et al. (1974) studied the effects of early protein-calorie malnutrition on a variety of problem-solving measures, measures which differed in the extent to which they placed a strong attentional demand on the subject. On tests which did not make strong attentional demands, monkeys reared on low-protein diets did not differ from well-nourished controls in performance. Previously malnourished monkeys, however, showed poorer performance on discrimination tasks, which required sustained attention. The authors concluded from their research that "animals suffering from protein-calorie malnutrition do not develop the supporting habits that lead to the successful solution of problems that require attention (p. 124)."

In interpreting the findings of Zimmerman and his colleagues, Riopelle (1974) has noted that the nutritional deprivation occurred well after the period of cell division, which ends at about three months postnatally, after the period of most rapid brain growth, and when myelinization—one of the last steps in brain development—was nearly completed. Riopelle suggested that this finding indicates that the effects of malnutrition on attention, motivation, and distractibility may be related to cerebral organization, that is, its functional organization, rather than to structural impairment. Perhaps, most importantly, he noted that the findings suggest that the behavioral impairments may be the result of a "more general deprivation syndrome" (Riopelle, 1974,p. 131), rather than the effects of protein-calorie deficit per se. In his words, "overt behavioral changes may be produced by internal changes that are common products of a variety of deprivations, among which may be felt hunger, persistent restlessness, unrelenting frustration, and sensed inability to succeed." These are important considerations and are related closely to the interpretive issues that arise when we consider human malnutrition.

3.3.4.1.1 Short-term effects Levitsky et al. (1979) have pointed out that it is important to distinguish the question of concurrent or short-term effects of malnutrition on animal learning from the long-term effects. Further, it is important to distinguish learning tasks as being simple (i.e., requiring a simple stimulus-response association) or complex (i.e., requiring multiple stimulus-response associations or complex problem-solving). Malnutrition is most likely to show concurrent rather than long-term effects on animal learning and to result in impaired performance on complex rather than simple tasks. Levitsky et al. (1979) have summarized the literature, making the following points.

The major problem in interpreting the results of studies utilizing simple learning tasks is separating the effects of the malnutrition on learning ability from the effects on motivation and performance. The two approaches most frequently used to investigate associative learning in malnourished animals are learning an association for food reinforcement and learning an escape response from an aversive stimulus. With food reward tasks, there have been conflicting findings. For example, a number of early studies (Bevan & Freeman, 1952; Riess & Block, 1942) showed that rats malnourished beginning at weaning learned a simple maze significantly faster under a food reward condition than control animals. Other researchers, however, found poorer learning under similar dietary restrictions (Rajalakshmi et al., 1965), and still others found no differences (Zimmerman et al., 1974).

A key variable in such studies is the reinforcement value of the food reinforcer. In the Zimmerman et al. study (1974), the investigators attempted to equate the reinforcement value of the food reward in the experimental and control groups. While there were effects of the concurrent malnutrition on performance in the learning situations (for example, attention to task demands), there was no evidence that the protein deprivation had impaired the ability to make simple discriminations, learn a reversal operation, acquire a learning set, or learn a delayed response.

A problem with either escape or avoidance learning tasks is the difficulty in separating response to the task learning requirements from response to the aversive stimulus per se. There is evidence (reviewed earlier) that malnutrition has significant effects on emotional control and, in particular, response to moderately stressful stimuli. Thus, poor performance on tasks involving such stimuli may reflect over-reactivity to the stressful stimuli, rather than inability to learn the task. For example, Cowley and Griesel (1962) found that young rats, malnourished from conception, performed more poorly than controls when trying to escape the water in a water maze task only when water temperature was below room temperature. When water temperature was at room temperature, the malnourished rats performed better than the control rats. The earlier-cited study by Barnes et al. (1966), which showed poorer learning in the malnourished rats, also was conducted in unusually cold water. Levitsky concludes, "there exists no evidence that concurrent calorie or protein malnutrition introduced early in life can clearly affect S-R learning independent of the potent effects calorie and protein malnutrition have on performance in simple learning situations." (Levitsky et al., 1979, p. 277)

The rationale for examining learning on complex tasks or tasks of "animal reasoning" is that rate of associative learning is only one aspect of adaptive intelligence. Often an animal is faced with a higher-level problem-solving task, in which alternative responses are available and the best must be chosen, or in which a number of consecutive or simultaneous associations or discriminations must be made in order to achieve successful solution. In short, it may be that malnutrition differentially affects the ability to learn simple associative responses and the ability to utilize available responses to solve more complex problems.

A test procedure developed to look at more complex learning is the Hebb-Williams maze, developed by Hebb and Williams (1946) and modified by Rabinowitz and Rosvale (1951). In this task, the animals are placed in a large box. After the animals are familiarized with the maze, a series of barriers is constructed between the start and goal boxes. The dependent measure in this task usually is the number of errors made by the animal in negotiating the maze.

In general, malnourished animals perform more poorly on this task than control animals when malnutrition is prenatal (Cowley & Griesel, 1959, 1963, 1966), at parturition (Baird et al., 1971; Zimmerman & Wells, 1971), or post-weaning (Baird et al., 1971), as long as the malnutrition is continued through the time of testing.

It is likely that motivational and attentional differences underlie the impoverished performance. Strupp and Levitsky (1983) have distinguished between "essential learning" and "advantageous learning." Advantageous learning is defined as the "acquisition of information that is not motivated by an immediate biological need nor is overtly reinforced" (p. 536); that is, the learning of nonessential information or information which, although not necessary for survival, may give the learner an advantage over nonlearners in certain future problem-solving situations. The classic example of advantageous learning is what human researchers call "incidental learning," the learning of features of a task which are not essential for performance but which may facilitate future performance on related tasks.

The findings by Zimmerman et al. (1974) on complex problem-solving by monkeys and the findings of the many investigators who have examined malnourished animals' performance on the Hebb-Williams task suggest that it may be the ability to attend to and learn nonessential information which is most affected by malnutrition (Levitsky & Strupp, 1984). (Note that the studies of food reward learning and avoidance or escape learning are tests of essential learning.) Concurrent malnutrition has been shown to influence redundant learning (the learning of information superfluous to a task because it is redundant), latent learning (incidental learning), and observational learning (learning by observation of a model without opportunity for performance) in animals (Levitsky, 1979; Strupp, 1982; Strupp & Levitsky, 1983). This distinction between advantageous learning and essential learning is critical in our understanding of the behavioral effects of malnutrition because it helps to account for important findings on the behavioral effects of undernutrition in humans.

3.3.4.1.2 Long-term effects Results of studies on the long-term effects of early malnutrition on later animal learning are not consistent. The results are highly dependent on whether or not the malnutrition has been accompanied by environmental deprivation and also on specific task characteristics.

As with studies on concurrent effects of malnutrition on behavior, there are invariable confounds when food is used as a reward in a learning task. The problem is that a history of malnutrition is likely to influence the incentive value of food for the malnourished animal (Levitsky et al., 1979). Specifically, malnutrition introduced during lactation or shortly after weaning increases the motivation for food, both in the previously malnourished adult (Mandler, 1958; McKelvey & Marx, 1951) and in the currently food-deprived adult (Smart et al., 1973). Since the same food reward may not be of equal incentive value to animals with different nutritional histories, one cannot necessarily attribute differences in performance on learning tasks to differences in learning ability.

Given the unknown effects of differences in motivation, the fact that the effects of different food reward incentive values on performance depend on the complexity of the learning task (Kimble, 1962) and the fact that associative learning tasks differ in complexity, it is not surprising that the results of findings often are conflicting.

A similar problem occurs when we consider learning paradigms which involve aversive stimuli (Levitsky et al., 1979). If there are long-term effects of malnutrition on emotional responsiveness (and there is such evidence, as shown in the previous section), then emotional and motivational factors may be implicated in the differences in performance between animals with and without histories of malnutrition on avoidance or escape learning tasks. It may be that the confounding of motivational and ability effects is responsible, in part, for the general lack of consensus in results of studies which have investigated long-term effects of malnutrition on animal learning.

The most thorough review of research on the long-term effects of early malnutrition on animal behavior is by Levine and Wiener (1976b). They review the studies on active avoidance learning, passive avoidance learning, and maze learning. Active avoidance learning paradigms require the animal to learn to avoid a noxious stimulus, usually an electric shock, by running or jumping. In passive avoidance learning, the animal avoids the stimulus by not responding. Maze learning paradigms have included the simple visual-discrimination mazes and the more complex Hebb-Williams mazes.

Studies on long-term effects of early malnutrition on animals' capacities to make an active avoidance response to an aversive stimulus yield conflicting results. For example, Caldwell and Churchill (1967), investigating the effects of malnutrition during the second half of gestation on the ability of adult rats to learn a pole jump avoidance task, found that prenatally malnourished rats (all subjects had been on high-protein diets after birth and dams had been given high-protein diets during lactation) learned the task—jumping onto a pole to avoid a shock occurring five seconds after a warning tone—but that their latency to avoidance response was longer than that of control animals. Frankova and Barnes (1968) examined effects of early postnatal malnutrition in rats on performance on a screen jump avoidance task. In contrast to the Caldwell and Churchill subjects, previously malnourished rats did not show a longer response latency than controls on the avoidance task. Frankova and Barnes reported differences between the two groups in the ability to *extinguish* the response. After several trials of warning tone followed by no electric shock, the previously malnourished animals were more likely to continue to make the avoidance response. Frankova and Barnes also noted the unusual "emotional" behavior of the malnourished animals during the intertrial waiting periods.

Several studies have, however, shown no differences between malnourished and nonmalnourished animals on tests of active avoidance. Studies by Guthrie (1968), Di Benedetta and Cioffa (1972), and Myslivecek, Hassmanova, and Springer (1969) found no long-term effects of early postnatal malnutrition (based on large litters) on the ability of rats to learn an active avoidance task later in life. Two studies have suggested *positive* effects of the malnutrition. Morris (1974) compared rats that had been raised by mothers receiving a low-protein diet until weaning with rats that had been given a high-protein diet throughout the period of lactation. When tested at 13 weeks of age, the previously malnourished group made more active avoidance responses per session than the well-nourished group. And Frankova (1973a), in a follow-up study to the Frankova and Barnes study cited above, replicated their finding that rats malnourished both prior to weaning and following weaning (to 7 weeks) differed from nonmalnourished rats in terms of emotional behavior on the avoidance task, but did not differ in response latency to the conditioned stimulus on the screen jump avoidance task. Further, they actually learned the task faster than nonmalnourished animals; they needed fewer trials to obtain error-free performance. Interestingly, a change in the methodology which familiarized the rats with the test situation prior to testing had the effect of eliminating the differences in emotional behaviors (so that there were no differences in frequency of abnormal jumping behaviors) but resulted in a difference in speed of learning the avoidance task, with malnourished animals now showing poorer response acquisition.

Passive avoidance experiments have tended to show that previously malnourished animals perform better than animals without a history of undernutrition. Levine and Wiener conclude from their review of the studies that the results argue against the position that significant protein-calorie deficits pre- and/or postweaning necessarily impair learning abilities in adulthood. If we consider the studies in which such higher performance of malnourished animals has been found, we see, however, that one could as easily interpret the poorer perfomance of better-nourished animals as showing their higher motivation and persistence and, in particular, exploration and interest in the environment.

Major studies in this area have been conducted by Hanson and Simonson (1971), Rider and Simonson (1973, 1974), Levitsky and Barnes (1970), Smart and Dobbing (1972), and Smart et al. (1973). Simonson and colleagues examined the responses of rats to a situation in which they had been deprived of water overnight and then were allowed access to a drinking chamber from a nondrinking chamber. Malnourished rats were prenatally malnourished, with mothers of these rats fed diet-restricted or low-protein diets during gestation. Experiments were conducted when the rats were adults. The experiment began with the rat being placed in a nondrinking start chamber for two minutes. Then a gate was opened, and the rat was allowed access to water. After a number of no-shock drinking trials, a single shock was delivered to the water when the rat licked the water. Latency to drinking (first lick) was measured on both preshock trials and on the subsequent shock trials. Malnourished rats were slower to cross into the drinking chamber initially than were control rats. After the shocks began, the difference in response latency for licking the water between malnourished and control rats increased. That is, malnourished animals now were performing relatively better (i.e., responding more slowly) than the control animals on the avoidance task. Finally, when the shock trials ended, the malnourished animals did not show the same return to baseline level of responding. These results were obtained both for animals in which dams had been diet-restricted (to 50 percent of normal diet; Hanson & Simonson, 1971) and protein-restricted (Rider & Simonson, 1974), but not in animals whose mothers had been provided high-protein, low-calorie diets during the perinatal period (Rider & Simonson, 1973). The findings with regard to preshock and shock trial performance suggest that the protein malnourished animals were less motivated to explore the environment and less persistent in meeting their physical needs. The postshock results suggest attentional deficits.

In the study by Levitsky and Barnes, subjects were placed on an elevated platform above an electrified grid. They were permitted to step down on the grid, where they then were shocked. Rats that had been nutritionally deprived prenatally with mothers fed a low-protein diet , postnatally by protein restriction, or both, showed a longer latency to step down onto the grid than control animals on *postshock* performance. While Levine and Wiener interpret this difference as reflecting better performance in the malnourished animals, it should be noted that the longer latencies in the *postshock* period in the malnourished animals were no longer functional; there was no purpose served by the animals taking longer to respond to the opportunity to step down on the grid. The results indicate not an improved learning in malnourished animals (of a passive avoidance response) but a *failure to disinhibit an exploratory response*. This could reflect either a weaker motive to explore (a questionable interpretation, given the lack of behavioral differences on preshock or shock trials) or a failure to learn an appropriate functional response—due, perhaps, to a failure in attention or a lack of interest in the problem.

Finally, the studies by Smart and colleagues involved a two-chamber apparatus in which the animal was placed in a starting chamber and 10 seconds later was given an opportunity—by virtue of a sliding door being opened—to enter a second chamber. The animal was shocked when it entered the second chamber. A study of offspring of diet-restricted mothers (50 percent of normal intake during gestation and lactation) who were habituated during weaning showed no differences at 30 days of age with respect to latencies to cross into the second chamber on either the first or second shock trials (Smart & Dobbing, 1972). A follow-up study of the rats in adulthood (Smart et al., 1973), however, showed that the original experimental group—who had been both prenatally and postnatally protein-calorie deprived—and also an experimental group that had been only postnatally deprived (i.e., during lactation) showed signficantly different performances than control animals or those only prenatally deprived on the passive avoidance task. Specifically, the postnatally deprived animals showed significantly longer response latencies on both the first and second trials. The results, consistent with those of the Simonson group and Levitsky and Barnes, indicate both poorer attention and/or interest in the physical environment. It should be noted that the malnourished group did not show a significantly greater increase in latency from trial one to trial two than the control group. Thus, there is no evidence that the malnourished group had better learning of the stimulus-response contingencies.

Interestingly, Levine and Wiener conclude their review of passive learning studies by remarking, "It appears that the results of these studies agree that the prenatally malnourished rat exhibits a better passive avoidance behavior" (1976b, p. 128). They also note, with regard to the consistent finding that previously malnourished rats fail to unlearn the passive avoidance response as readily as nonmalnourished animals, that "Resistance to extinction frequently has been used as an indication of greater learning" (p. 128). However, failure to unlearn a nonfunctional response so that an alternate (e.g., exploratory) response can be adopted can be interpreted as evidence of impaired behavioral functioning. For example, as we will see later, Lester's (1975) finding that malnourished human infants could not *dishabituate* an orienting response —that is, stop orienting to a new stimulus when the stimulus became familiar—has been interpreted as evidence that malnutrition adversely affects the ability of the organism to organize itself for increased interaction with the environment (see Brazelton et al., 1977).

A final category of studies is maze learning. The majority of studies examining long-term effects of early malnutrition on maze learning have examined complex maze performance, generally the Hebb-Williams maze. Both Levine and Wiener and Levitsky et al. agree that results have varied as a function of time of prior malnutrition (e.g., gestation vs. lactation vs. postweaning), duration of malnutrition, and type of deprivation (i.e., the specific combination of deprivation conditions). For example, in most studies in which rats were malnourished during gestation and malnutrition was continued throughout lactation, malnourished rats performed more poorly than control rats as adults on the Hebb-Williams maze (Cowley & Griesel, 1966; Ottinger & Tanabe, 1969; Randt & Derby, 1973; Tanabe, 1972; Vore & Ottinger, 1970). In one study in which malnutrition extended through lactation and into the postweaning period, no differences in adult performance between malnourished and control rats was discerned (Baird et al., 1971). Studies in which malnutrition occurred only during gestation have shown poorer performance in malnourished rats (Cowley & Griesel, 1966; Vore & Ottinger, 1970) as well as no differences in performance (Ottinger & Tanabe, 1969; Smart et al., 1973). Two studies involving malnutrition during lactation only have shown superior performance in malnourished rats compared to controls (Slob et al., 1973; Smart et al., 1973). Yet, two other studies employing malnutrition during lactation only showed poorer performance on the Hebb-Williams maze in the malnourished rats (Ottinger & Tanabe, 1969; Vore & Ottinger, 1970).

The conflicting results are difficult to interpret, and it appears that we must go beyond a pure learning interpretation to make sense of them. A key may lie in the observation that performance on the Hebb-Williams maze is highly sensitive to attentional, motivational, and emotional factors (Stein, 1974). As shown in the previous section, a striking short-term effect of malnutrition on animal functioning is an increase in emotionality and, specifically, increased sensitivity to particular environmental stresses or frustrations. Depending on the task, different degrees of nutritional deprivation may result in better or poorer performance on a task such as the Hebb-Williams task, depending on whether the heightened reactivity increases performance incentive, inhibits desire to perform, or has no effect due to the particular task variables.

3.3.5 Consistencies in findings

Research on the effects of malnutrition on exploration, emotion, social behavior, and learning is linked by the construct of coping with stressful situations. Undernourished animals have difficulty making adaptive responses to difficult or stressful situations. Thus, in situations in which healthy animals inhibit emotional responses to make an adaptive response, malnourished animals have difficulty. In highly novel situations, malnourished animals fail to explore, showing either inhibition (i.e., under-responsiveness) or aimless activity. Malnourished animals frequently fail to make normal social responses to agemates, again presumably because of their difficulty in dealing with situations in which there is some stress (e.g., uncertainty about outcome). And in learning situations in which there are strong competing responses—that is, in which it is easier to make an alternative response—and in which learning is advantageous but not necessary to the organism, performance is likely to be decreased.

The inability to make the adaptive coping response under stress—to inhibit the overlearned responses that prevent the making of more adaptive responses—is the major outcome of undernutrition in animals. Thus, the undernourished animal fails to explore the new environment, inhibit maladaptive responses in avoidance situations, learn new responses that depend on attention to incidental stimuli, and learn to respond to social initiations from other animals.

In sum, the malnourished animal withdraws from interaction with the social and nonsocial environment, "functionally isolating" itself (Levitsky & Barnes, 1972) from learning experiences. Over time, the animal begins to have difficulty coping with situations that require novel responses, accommodation to new environmental demands, while the healthy animal learns more adaptive coping responses.

3.3.6 Possible bases for inconsistencies in findings

There is disagreement about the effects of malnutrition on learning capacity, the ability to form associations and educe relations. Our review of concurrent and long-term effects of malnutrition on learning does not allow us to draw a simple conclusion about the effects of malnutrition on the capacity to learn. The effects of malnutrition on performance on tests of learning depend largely on the extent to which the animal must attend to incidental stimuli, the degree to which learning is essential to survival and/or is positively rewarded, the nature of the reinforcement for learning, and the degree to which the animal must inhibit a well-learned response.

Levitsky (1979) states that most tests of simple associative learning do not show long-term or even concurrent effects of malnutrition on learning per se, independent of the above performance factors. However, most tasks *do* involve these different performance factors. The inconsistency in findings is due in large part to the fact that in a given task several of these factors may operate simultaneously and differ in the extent to which they "favor" the malnourished or well-nourished animal. Thus in a task in which attention is focused and in which there is a high probability of the desired response (i.e., there is no high-probability "competing response"), the malnourished animal is at a comparative advantage, since attentional demands and the need for control of emotional response are minimized. Yet in the same task under slightly different conditions, differences between well-nourished and malnourished may emerge. Thus, we have, for example, the significant effects of malnutrition on maze learning under cold water conditions but not under conditions of normal water temperature.

It may be that the learning/performance distinction is less important in practice than in theory. Learning is inferred from performance. The ability to perform in learning tasks is dependent on the animal's capacity to attend and to inhibit irrelevant or interfering responses, as well as to form associations and execute motor responses. The fact is that most, although not all, studies of the concurrent and long-term effects of malnutrition on performance in learning situations show adverse outcomes (Halas & Sandstead, 1980) due, perhaps in large part, to the adverse effects of malnutrition on attention and motivation. While many studies do not show adverse effects, the preponderance of data suggest that the effects of malnutrition on the animal's ability to adapt to new environmental demands, to inhibit maladaptive responses and to attend to task demands under stress, ultimately influences the animal's ability to perform in a variety of learning situations.

3.3.7 Research on the interaction of malnutrition and environmental deprivation

One of the most important developments in research on malnutrition and behavior has been the recognition on the part of scientists that there are striking similarities between the effects of malnutrition on behavior and the effects of environmental deprivation and, further, that the effects of malnutrition may depend on the level of environmental stimulation or deprivation.

Effects of sensory stimulation on central nervous system development have been well documented. In general, environmental stimulation leads to an increase in brain weight and size (Walsh & Cummins, 1976). Other biochemical and structural parameters affected are dendritic arborization (Walsh & Cummins, 1975), size and number of synapses (Cragg, 1974), and amount of protein and RNA (Rosenzweig et al., 1972). Animal research also has shown that environmental stimulation can reverse or compensate for the effects of deprivation. This has been demonstrated for the anatomical parameters of cerebral weight and size. Walsh and Cummins (1976) constructed four groups of rats. One large group had been reared in complexity for 535 days, and one group had been reared in isolation for the same period of time. One-half of each of these larger groups was tested on the Hebb-Williams maze (i.e., as a form of environmental enrichment) and one was not tested. For isolates, testing raised cerebral values for weight to about the same level as complexity-raised animals. Complexity-raised animals were not affected by the additional stimulation. Isolation-reared animals who were not exposed to testing showed significantly lower cerebral weights than the other groups.

The same investigators also have shown that stimulation not only can enhance recovery, it can protect against the consequences of isolation. In another four-group study, rats were exposed to 60 days of isolation, 60 days of complexity, 30 days of isolation followed by 30 days of complexity, or 30 days of complexity followed by 30 days of isolation. Animals in the switchover groups had cerebral weights as large as animals in the continuous-complexity group, and significantly larger than animals experiencing isolation throughout. The anothers noted that the experiment indicated "not only that complexity was sufficient to overcome the effects of prior deprivation, but also that earlier exposure to complexity afforded protection against subsequent isolation." (p. 173)

In addition, behavioral effects of early stimulation and deprivation have been identified. Denenberg (1964) demonstrated that early handling in animals results in lowered emotionality as evidenced by open-field behavior. Harlow et al. (1968) and Rosenzweig et al. (1972) showed improvements in problem-solving following early stimulation. Levine et al. (1956) showed improved performance on an avoidance learning task after early handling. In contrast, studies of severe environmental deprivation have shown long-term effects of aversion to novel stimuli (Mason & Green, 1962), fearfulness (Mitchell & Clark, 1968), reduced exploration and impaired problem solving (Ganz, 1968; Harlow et al., 1968).

Integrating these findings with those summarized earlier on the behavioral effects of malnutrition, it can be seen that the effects of early malnutrition on animal growth and behavior often are the reverse of those obtained when animals are exposed to increased environmental stimulation. Whereas central nervous system structural development is retarded by early malnutrition, early stimulation (i.e., increased handling) appears to accelerate the development of the central nervous system. Also, the increased emotionality associated with malnutrition, the decrease in exploration, and the poorer performance on certain discrimination tasks contrasts with the lowered emotionality, higher exploratory activity, and improved discrimination learning in early stimulated animals. Such observations have led investigators to consider the possibility that the effects of malnutrition may be moderated by social-environmental factors and, further, that they may be reversed by later stimulation.

This issue has been addressed in the research of Frankova (1974), Levitsky and Barnes (1972), and Elias and Samonds (1977). Frankova attempted to increase social stimulation for infant rats by rearing them with an "aunt" rat—a nonpregnant virgin female. In general, the result of introducing this new member into the mother-infant family was increased interaction between individual pups and more interaction between the rat pups and the mother in the presence of the aunt. Most importantly, behaviors of protein-deprived infant rats who were raised with the aunt rat were markedly different than those of protein-deprived rats raised only by the mother. Exploratory activities were enhanced, there were higher levels of social interaction among pups, and, in general, symptoms of abnormal behavior associated with malnutrition declined (see Frankova, 1973a).

The interactive effects of malnutrition and environmental stimulation were apparent through adult life. Rats in the four groups were compared for maternal behaviors with their own pups and with alien pups (Frankova, 1981). With their own pups at ages 6 and 8 days, protein-deprived rats reared without the aunt showed a longer latency to contact with offspring and made fewer contacts than rats from the other groups. They also scored lower on a scale for behavior in the nest. Protein-deprived rats reared by the aunt had lower response latencies, longer contacts, and higher nest scores than malnourished rats reared without the aunt. Similar effects occurred for behaviors with alien pups of 10 and 11 days. Protein-deprived rats not reared by the aunt had a longer latency to first approach to the pups, a longer latency to first contact, briefer contacts with the pups, fewer contacts, and more negative behaviors (e.g., throwing or pushing a pup away). Protein-deprived rats reared by the aunt showed better maternal behavior on each of these measures, performing more like well-nourished rats. In sum, there were significant effects of protein deprivation on maternal behavior of the rats, but they were moderated by early experience. To some degree, an improved maternal environment compensated for the effects of the early malnutrition.

Levitsky and Barnes (1972) examined the effects of malnutrition during the first seven weeks of life on the social behaviors of rats raised in stimulated, isolated, or control environments. The stimulated environments condition provided for extra handling of rat pups before and after weaning, and a one-hour play period with other rats and with toys five days a week during the four-week postweaning period. The isolated environment condition was characterized by a minimum of handling prior to weaning. Following weaning, animals were placed in closed wooden chambers and received no handling. At seven weeks, all animals were placed in control environments and provided with control diets. Following ten weeks of recovery, the investigators obtained measures of locomotor activity in the open field, following behavior, fighting, and exploratory behavior. On all measures except fighting, effects of malnutrition were increased by environmental isolation and decreased by environmental stimulation. For example, while malnourished rats showed less exploratory behavior than well-nourished rats, the difference was significant only for rats raised in the isolated environments. Similarly, the tendency for malnourished animals to show less following behavior than well-nourished animals, evident in the control and isolated environment conditions, was not apparent in the stimulated environment condition. In the case of fighting, however, while malnourished animals generally were more aggressive than well-nourished animals, this effect was diminished in the isolated environment; in the isolated environment, there were no differences between the two groups in the frequency of fighting.

Elias and Samonds examined the growth and development of 32 cebus monkeys, following environmental deprivation and/or food restriction. Since, in their words, exploratory behavior is "the aspect of behavior most affected by malnutrition" (Elias & Samonds, 1977, p. 358), exploration was the primary dependent variable.

The period of insult was two to six months, and was followed, in all cases, by six months of rehabilitation. Subjects were assigned to either isolated rearing or enriched rearing conditions. Subjects were calorie restricted, protein restricted, protein-calorie restricted, or unrestricted. Testing was performed during 17 weekly observations from 2 months to 6-1/2 months of age.

Both diet and rearing produced effects on behavior. Calorie deficiency produced an effect on behavior independent of rearing conditions, with calorie-restricted animals showing less exploration than control animals, even with enriched rearing. Protein restriction produced deficits in exploration under conditions of isolated rearing but not under enriched rearing. Protein-calorie restriction produced behavior impairments under both isolated and enriched rearing conditions.

Barnes (1976) has suggested that two effects of the apathy and reduced curiosity associated with early malnutrition are a "functional deprivation of experiences that are important to cognitive development, and a functional avoidance of new experiences and novel stimuli." The functional deprivation/avoidance hypothesis implies a circular process: Adapting a passive, self-protective style of responding to new situations and increased environmental demands leads to a reduction in the capacity to respond appropriately to such events, and a tendency to perform inadequately in situations which require attention to incidental stimuli, persistence, control of emotions, or motor inhibition. Results of the studies cited above suggest that such a cycle may be interrupted by an intensive and sustained program of environmental stimulation.

Walsh and Cummins view the issue from a similar perspective. They suggest that "the nature and amount of sensory stimulation may set a limit to the function and development of the brain," and that "This (sensory stimulation) is only one example of a large range of other factors (genetic, nutritional, hormonal, etc.) any one of which may determine the limits of functioning in the normal brain or may limit recovery after brain damage" (Walsh & Cummins, 1976, p. 187). They propose that the functional ceiling will be determined by the most deficient factor, which they suggest be called the Function Limiting Factor.

What might be the basis of the facilitating effects of environmental enrichment and/or the adverse effects of deprivation? The authors suggest the role of environmental feedback and its effect on the behavior of the organism. For any type of environmental enrichment to have a beneficial effect on behavior, the organism must interact with the stimuli and reafferent stimulation must occur. This means that the organism creates stimulation through its increased interaction with the environment, stimulation that reinforces the behavior which elicited the stimulation.

When we consider the functional isolation hypothesis of Barnes and Levitsky, the relationship to the findings on malnutrition and behavior is clear. Malnutrition and environmental deprivation both have the effect of functionally isolating the animal from the physical and social environment. With improved nutrition or stimulation, the animal increases interaction with the environment and increases the level of stimulation, reinforcing those behaviors which led to the increased contact. According to this reinforcement framework, the effects of environmental stimulation (including nutritional intervention) on behavioral recovery are mediated not only by direct physiological effects on the nervous system but by the eliciting and reinforcing of therapeutic behavior (Walsh & Cummins, 1976, p. 190).

4.0 HUMAN STUDIES ON MALNUTRITION AND BEHAVIOR

4.1 Organization of the Review of the Studies

In reviewing the human studies on malnutrition and behavior, we can classify studies in terms of the type of design and the population considered. There are several different categories of studies with respect to design and population. First, and most prevalent, have been studies comparing children who earlier in life were hospitalized or treated for clinical protein-energy malnutrition. A second category of studies compares children who, earlier in life, experienced protein-energy deprivation as secondary to neonatal illness and, in particular, intestinal disorder. A third category of studies examines the effects of nutritional supplementation (for mothers and/or infants) on the later behavior of children from endemically malnourished populations. Fourth are studies of the behavioral characteristics of children diagnosed as having nonorganic failure-to-thrive syndrome, a medical syndrome diagnosed in early infancy where the child does not grow but there is no known organic cause so that nutritional and social-interactive variables are assumed to be implicated. A fifth category of studies is those examining the behavioral sequelae of intrauterine growth retardation as evidenced by depressed weight for length at birth or depressed weight for gestational age. Sixth, there are a few studies that have attempted to examine the relationship between chronic undernutrition and later behavior by examining correlations between anthropometric measures and later behavioral outcomes in chronically deprived populations. To the extent that the anthropometric measures reflect nutritional history (i.e., individual differences in nutritional history), associations between chronic malnutrition and behavior can be inferred. A final category includes clinical intervention studies of severely malnourished infants.

The different types of studies address different research issues. For example, the failure-to-thrive studies concern the effects of growth failure in infancy when there are assumed to be (i.e., already) problems in mother-child interaction and/or feeding disturbances which contribute to growth failure. In contrast, the studies of children with and without histories of malnutrition (both the studies of children who suffered from primary protein-calorie deprivation and the studies of children who were malnourished secondary to neonatal illness) attempt to control as strongly as possible for home-environmental differences, including intrafamilial variables. Thus, the studies address different empirical questions and do so from different theoretical perspectives. Yet, together, the studies, when examined critically and recognizing their different methods and theoretical frameworks, contribute to our understanding of the long-range sequelae of early protein-calorie malnutrition.

In the present section we discuss the findings from the different types of studies on human malnutrition and child behavior. At the beginning of each section we discuss the methodological features of each type of study, and when possible, identify the general class of research design (as defined in Chapter Two) it represents.

4.2 Retrospective Studies of Severely Malnourished Children (Comparative, Prospective Design with Data Collected Retrospectively)

4.2.1 Design characteristics

Most of the studies on the effects of severe malnutrition on behavior use nonexperimental designs and are prospective from the point of view of data analysis. That is, children are chosen for the study because of their standing on a presumed antecedent variable and then compared on the dependent variable(s). In the case of malnutrition research, children are identified who are known to have suffered from early and severe malnutrition and they are compared at later ages (early childhood, later childhood, adolescence) on different developmental and behavioral measures.

There are points to be noted about this type of study. First, the studies are retrospective in data collection in that the children often are first seen at a later period in time and then studied with respect to their earlier nutritional history. But the analysis is prospective, since the selection variable is the presumed antecedent variable (i.e., malnutrition) and not the developmental outcome. Second, different studies may include components that represent other types of designs. For example, following the first outcome assessment, children may be followed longitudinally. Finally, the specifics of design, including construction of control groups and methods for matching on background characteristics, will differ among the different studies. For example, Hertzig et al. (1972) used a classmate control group and a nonsibling control; Galler et al. (1983a) used a sibling control group only, and Monckeberg (1968) used a nonsibling control.

Designs falling into this category are potentially vulnerable to all of the threats to internal validity described in Chapter Two. In addition, the construct validity of inferences is a concern since the condition conceptualized as undernutrition or hospitalization for undernutrition may represent a much broader kind of environmental deprivation. For these reasons some researchers have concluded that such designs are inherently inadequate (Warren, 1973). As shown earlier, however, the inherent interrelationships between nutritional and extranutritional variables also can be taken advantage of as we design, carry out, and interpret such studies. In particular, such studies provide information about the risk likely to be attached to the experience of severe malnutrition and hospitalization—in the social-environmental context in which it normally occurs—with regard to the child's later functioning.

Perhaps the most underestimated threat to the construct validity of the retrospective studies is the possible confounding of current undernutrition with early undernutrition and of micronutrient deficiencies with gross energy-protein deficit. This is a concern that has only recently been articulated (Frank, 1984). Research that enables us to better separate the acute components from the chronic components of undernutrition, as the Hoorweg (1976) study attempted to do, makes a major contribution to this field. Studies comparing the effects of long-term undernutrition with and without current undernutrition superimposed have obvious social policy implications. This issue is discussed further in Chapter Five.

4.2.2 Review of research

4.2.2.1 Effects on infant behavior and development Of the studies investigating the effects of clinical malnutrition on infant behavior, most have examined performance on standard developmental schedules. These tests, for example, the Bayley and the Gesell, involve an assessment of a broad range of functions including motor, perceptual motor, social, and rudimentary language, as well as cognitive. Two studies have looked at specific behavioral competencies: Lester's studies on orientation and Brockman and Ricciuti's studies on concept development.

Lester (1975) examined the cardiac-orienting response, a measure of central nervous system integrity, in 1-year-old Guatemalan infants. The index group was 55–75 percent of average weight for age, while the control group was of average weight. Lester administered pure tone stimuli to the infants and used heart rate deceleration as a measure of the orienting response. While the well-nourished infants showed the expected heart-rate deceleration on the first trial after a new tone was presented, the poorly nourished infants did not show any effects over trials: There was a diminished orienting response, no habituation over trials, and no dishabituation following the introduction of a new tone. The results indicated an attentional impairment.

Brockman and Ricciuti (1971) examined the cognitive functioning of infants hospitalized for nutritional marasmus. Subjects were Peruvian children, tested at 1–2 or 3–4 years of age. The 20 experimental subjects had been diagnosed as marasmic, with body weight less than 50 percent of expected weight for age. Control subjects were matched for age and sex with the patients, but had body lengths at or above the Boston growth curves. There were significant differences in categorization behavior between experimental and control children, based on a locally developed object-sorting task. Differences favored control children at both the younger and older age levels.

One other study examined specific cognitive abilities in infants who had been severely malnourished. Cravioto and DeLicardie (1972) studied 22 Mexican children who had suffered from severe clinical malnutrition. Children were 4 to 53 months at the time of diagnosis. A total of 19 of the children were compared with children from the same birth cohort who did not develop clinical severe malnutrition, for language development, and for concept formation. The variables considered were (1) performance on the Gesell subscale for language development, and (2) performance on a test of bipolar concepts, such as big-little and long-short. Differences in language development in the experimental and control subjects were significant by 1 year of age, and they increased in magnitude between 1 and 3 years. Control subjects showed signficantly higher scores on the test of bipolar concept formation beginning at about 2 years of age.

The results of the Cravioto and DeLicardie study are difficult to interpret because subjects developed clinical malnutrition at different times—sometimes before a particular number of language tests were given, sometimes after. Thus, one cannot determine to what extent language differences preceded or anteceded the malnutrition. Further, as in the Brockman and Ricciuti study, the abilities investigated in the Mexico study had not been explicitly theoretically linked to malnutrition. Thus, although significant differences were obtained, one must speculate about the cognitive processes mediating group differences. As Riccuiti (1981) has noted, what have been broadly referred to as motivational and performance factors may have played a critical role.

The majority of studies examining effects of clinical undernutrition on infant development have used standardized tests of infant development. Tests used include the Bayley Scales of Mental and Motor Development, the Gesell scales, and the Yale Revised Developmental Schedules.

Tests of infant development do not measure the same aspects of functioning as intelligence tests, and they are not equally predictive of later intellectual and social functioning. They are suited primarily for developmental evaluation: Determining the extent to which the child has acquired those perceptual, motor, sensorimotor, and language skills which children typically acquire as they mature. It is important to recognize this explicit developmental focus. While all intelligence tests provide a comparison of a given child with the typical child of his/her age level, the age gradings for older children's tests are not as finely drawn as they are with infant tests. Specialists in infant development are able to speci[illegible] some degree of precision what a typical child of a given a[illegible]

13 months—should and should not be able to do. Such finely differentiated age gradations are not possible after the preschool years. Furthermore, infant tests focus on the acquisition of specific skills, and scoring generally is for presence or absence of specific skills. This is in contrast to standard intelligence tests, which provide not a description of specific abilities and limitations but, rather, a description of *relative* strengths and weaknesses (i.e., indicating presumed ease of learning or difficulty of learning) in specific areas.

Pollitt and Granoff (1967) studied 19 children who had been hospitalized between 6 and 8 months of age for nutritional marasmus (weight defict of equal to or greater than 40 percent) and compared them with siblings with no history of malnutrition. Subjects were Peruvian children, living in the barrios of Lima. At testing, within 24 months of hospital admission, experimental subjects showed significantly lower performance on the Bayley Scales of Mental Development than control children. There were no differences on motor performance.

One other study of infant behavior of marasmic children was carried out in Lebanon with subjects hospitalized between 2 and 16 months of age for acute, severe marasmus (Yatkin & Mclaren, 1970). In this study, however, there was no control group, and the design called for a longitudinal evaluation of the children for the 14 weeks after admission. Additionally, this study involved a quasiexperimental manipulation (stimulation). Results are discussed later in this chapter, along with other findings from studies examining the effects of nutritional and stimulation interventions on development.

A number of studies have examined the behaviors during infancy of children hospitalized for kwashiorkor. Of the studies focusing on infant assessments, however, none has employed matched control groups. These studies differ, then, from the other studies reviewed in this section. They are longitudinal studies in which subjects are their own controls.

In a clinical study described earlier, Geber and Dean (1956) administered the Gesell developmental schedules to African infants 1 to 3 years of age, who had been hospitalized for kwashiorkor. The authors reported that the children scored "Moderately" at the time of hospital release. As previously noted, perhaps as important as the results of the developmental evaluations were the clinical observations of the infants, with subjects described as anorexic, apathetic, and withdrawn. These observations have been corroborated by others, including Jelliffe (1965) and Latham (1968), both describing Jamaican infants. Summarizing the observations of the infant with kwashiorkor, Latham (1974) writes:

> Apathy, irritability, and a lack of interest and exploratory desire are features well known to those who treat children suffering from kwashiorkor. The child is usually miserable. He, or she, may lie almost motionless by the hour. . . . In successful treatment among the first changes to be noticed is an alteration in the mood of the child. . . . The first smile is evidence of an improved prognosis.
>
> (p. 545)

Cravioto and Robles (1965) and Barreda-Moncada (1963) studied the effects of kwashiorkor by examining the recovery of infants hospitalized for kwashiorkor. Subjects in the Cravioto and Robles study were 20 Mexican children, hospitalized for kwashiorkor between 6 and 30 months of age. A major purpose of the study was to determine the relationship between time of onset of illness and degree and rate of recovery. Subjects were divided into those admitted for hospitalization before 6 months of age, between 15 and 29 months, and between 37 and 42 months. Gesell tests were administered every two weeks following hospitalization for each group. Results showed that children hospitalized at the earliest period for kwashiorkor showed less complete improvement (and less rapid improvement) than subjects admitted at the later ages, with the children admitted at 37–42 months showing the greatest improvements. Since normative data on the Gesell are not available for the population, it is not possible to determine the extent to which the undernourished children would have differed from normal children of the same age. Moreover, although the results appear to indicate a deleterious effect of malnutrition on development as a function of time of onset of malnutrition, time of onset is confounded with length of hospitalization, which itself could influence behavioral development. Other criticisms of the study concern the small sample size, lack of overlapping data points in the three groups, and lack of follow-up information (see Barrett & Radke-Yarrow, in press; Hertzig et al., 1972).

Barreda-Moncada studied 60 Venezuelan children hospitalized for kwashiorkor between 15 and 72 months of age. Gesell schedules were administered to infants and young children after admission, either 7–12 weeks after admission or approximately two years after admission. Those tested shortly after admission showed markedly low developmental quotients (DQs), with mean overall DQs of approximately 65 (compared to an expected score of 100). At testing two years following admission, language DQs showed the most marked retardation (50–60) and motor the least (74). The interpretability of this study, like that of Cravioto and Robles, is limited by the fact that Gesell normative data for the population are not available.

A third group of studies includes those focusing on infants with mixed or undefined types of severe, clinical malnutrition. These include the studies of Stoch and Smythe in South Africa (1963, 1967), Cravioto and DeLicardie in Mexico (1972), and Chase and Martin in the United States (1970).

The research of Stoch and Smythe is a longitudinal study begun in 1955 in Cape Town, South Africa. A total of 20 subjects was identified as being seriously undernourished. All were malnourished between the ages of 10 and 16 months. Follow-up studies began in infancy and have extended into adulthood (Stoch et al., 1982). Gesell tests were administered up until age 2 and the Merrill-Palmer Scales were administered after age 2 and through age 6. Beginning at age 2, index children showed significantly lower developmental quotients than control children. The findings of this study, however, have been questioned (cf. Scrimshaw & Gordon, 1967; Warren, 1973) on the grounds that control children experienced living conditions superior to those of the index children with respect to parent employment, family stability, and sanitary conditions.

Cravioto and DeLicardie (1972) studied language and concept development in Mexican infants who had experienced maramus or kwashiorkor in infancy. Results of their study were summarized earlier in the section on specific behavior impairments.

Chase and Martin (1970) identified 19 children who had suffered from serious malnutrition during the first year of life and compared them at age 3-1/2 years with controls matched for age, sex, race, and social background. The mean developmental quotient for children who had suffered from severe malnutrition was significantly lower than that for matched controls. The investigators also found that age of hospitalization was significantly and negatively correlated with later IQ for index children. This finding is in contrast to that of Cravioto and Robles (1965). Whereas Cravioto and Robles interpreted their findings as indicating that early malnutrition had more serious conseqences than malnutrition later in infancy, Chase and Martin interpreted their findings in terms of duration of illness prior to hospitalization. They suggested that the poorer performance of children entering the hospital later in infancy was due to the fact that they were subject to malnutrition for a longer period of time than those entering the hospital earlier.

To summarize, studies of the behavioral effects of clinical malnutrition on infants have indicated concurrent and short-term effects on attention and orientation, rate of development of sensorimotor abilities, and concept formation. Studies employing both matched control groups and sibling control groups, as well as "subjects as own control" designs, have shown such effects to be associated with marasmus, kwashiorkor, and mixed or undifferentiated conditions.

4.2.2.2 Social-environmental correlates of malnutrition Retrospective comparison studies may be comparing children who differ on social-environmental conditions which may have helped to precipitate the onset of clinical malnutrition. Several studies have documented differences between the home environments of infants suffering from clinical malnutrition or recovering from clinical malnutrition, and those of nonmalnourished controls. Studies which have documented such effects include those by Cravioto and DeLicardie, Stoch and Smythe, Chase and Martin, Richardson and colleagues and Christansen et al. (1974).

Cravioto and DeLicardie (1972) compared the index and control children in their study, described above, on sociocultural characteristics of the families, family economic status, and a measure of the family environment; specifically, the home stimulation provided the child by the mother. One sociocultural variable was related to child clinical malnutrition: Three times as many mothers from the control group listened regularly to the radio as mothers from the index group. No other characteristics of the parents—biological (e.g., height), cultural, or social—differentiated the two groups. The authors did find, however, that there were significant differences between the groups in maternal stimulation of the children. Children in the control group had significantly higher scores on the Caldwell HOME Inventory (Caldwell, 1976)—which measures such characteristics as emotional climate in the home, vocal stimulation, and frequency and stability of adult contact—than did children in the control group.

Stoch and Smythe also found significantly different home environments in children who experienced severe malnutrition in infancy, compared with matched controls. The index children came from families with lower mean weekly incomes and lower levels of education (Stoch et al., 1982). In addition, control families experienced superior living conditions with respect to parental employment, family stability, and sanitary conditions (Scrimshaw & Gordon, 1967).

Chase and Martin (1970), in the study reported above, matched index and control children for birth weight, age, sex, and SES. Home environment comparisons, however, showed numerous differences between the groups. The malnourished children had more siblings in the family and more infant siblings per family. Parental separation, alcohol-related problems, and, as indicated previously, unwanted pregnacies, were more frequent in the families of malnourished children.

One of the most careful studies on social-environmental correlates of malnutrition was carried out in Jamaica by Richardson and colleagues (Hertzig et al., 1972; Richardson, 1972, 1974, 1976). The study, also referred to later in the sections on malnutrition and intellectual development and malnutrition and social development, involved a comparison of 74 boys who had been severely malnourished during their first two years of life with matched controls from their communities. Matching was performed on social class, as well as age and cultural experience. Index and control children then were compared on variables concerning the parent or guardian's education and "capacity," economic and housing conditions, and the child's educational resources. On each

TABLE 3-1

Variables used in the study of the background histories of Jamaican index and comparison boys

Guardian Variables
Guardian's upbringing and education
Mother's general reproductive history
Mother's pregnancy with study child
Guardian's level of capacity
Guardian's contact with media
Guardian's human resources
Mortality Variable
Mortality among siblings
Economic and Housing Variables
Structure and condition of the house
Home furnishings and appliances
Amount of crowding in the house
Educational and Social Variables (pertaining to child)
Extent and diversity of child's social relationships
Intellectual stimulation
Caretaker's child rearing practices
Amount of schooling

From Richardson, S.A. (1976). The influence of severe malnutrition in infancy on the intelligence of children at school age: An ecological perspective. In: R.N. Walsh and W.T. Greenough (eds.), *Environments as Therapy for Brain Dysfunction*. London: Plenum, pp. 256–275. Reprinted by permission of the publisher.

of the variables shown in Table 3–1, the previously malnourished group was inferior.

The data are important because they suggest the nature of the intracommunity, interfamily differences that may distinguish families of children who become malnourished from those who do not. They are important for another reason. As will be discussed later, the effects of malnutrition on intellectual development which were identified in this study were most severe for the children from the most disadvantaged homes. Thus, not only did the malnourished children differ from comparison children in social and economic circumstances, differences in those circumstances determined the degree to which the malnutrition would have longstanding, adverse effects on the children. The identified interaction is consistent with the findings on malnutrition and environmental deprivation in subhumans, and is a consistently reported outcome in studies on children.

Christiansen et al. (1974) provide a synthesis. Family resources (income), maternal education, crowding, occupational instability, and increased family size all are related to malnutrition. Increased family size and lack of material resources lead to crowded living conditions and inadequate health conditions. Poor family resources make difficult the acquisition of food. Occupational instability is associated with both poor material resources and inadequate parenting; that is, inadequate care for the child. And maternal eduation is related to the mother's capacity to provide adequately for her child.

Christiansen et al. also present data to support their assertions. In a pilot study to the Bogota longitudinal intervention study (Herrera et al., 1980), the investigators compared the home characteristics of families where both of two same-sex siblings were well-nourished and both of two same-sex siblings were malnourished. They also correlated different social status variables with child nutritional status, as assessed by height and weight for age.

Well-nourished children scored higher than malnourished children on a large number of measures of socioeconomic status including monthly income, father's and mother's education, parents' literacy, and ownership of closet, radio, and sewing machine. On four measures of sanitary conditions, well-nourished families were superior. In addition, there was significantly less crowding, parents (both) more frequently read the newspaper, and mothers had higher aspirations for their children (both in terms of education and occupation) and gave significantly more attention and stimulation to the child: more teaching, praise and encouragement, and playing with the child.

The variables significantly related to child height and weight included family income and expenses, parent literacy (both parents), father's education and occupation level, and ownership of radio and closet. The measures of sanitary conditions, crowding, reading habits, and aspirations for the children all were significantly correlated with these two measures of child physical status.

Findings such as these suggest that in studies of children with natural histories of malnutrition, one cannot separate the effects of poor environment from those of nutritional deprivation in accounting for behavior differences. As will be shown later, however, the two type of factors can be separated statistically and experimentally, and estimates of their separate effects obtained. Even more importantly, the role of social-environmental factors in setting the *conditions* under which malnutrition may be most dangerous can be documented.

4.2.2.3 Effects on cognitive development The area of study which has received the most attention has been the long-term effects of malnutrition on the intellectual development of the young child. Controlled studies falling into this category are those by Stoch and Smythe (1963, 1967, 1976, 1982), Champakam et al. (1968), Birch et al. (1971), Hertzig et al. (1972), Galler et al. (1983a), McLaren et al. (1973), and Hanson et al. (1971).

Stoch and Smythe conducted a study of 20 lower-SES South African children, malnourished between the ages of 10 and 16 months. Gross retardation of intellectual performance was found in the undernourished group when the children were measured at later ages. When tested at school age, malnourished children scored significantly lower than matched controls on subtests of the South African Intelligence Test for vocabulary, pattern completion, blocks, and form board. As indicated earlier, the findings of this study have been questioned (Scrimshaw & Gordon, 1967) on the grounds that control children experienced superior living conditions to index children. In addition, Stoch and Smythe (1976) themselves noted that their data are subject to question since control children were not adequately matched with index children on EEG readings. These ambiguities limit the interpretability of the study.

Champakam, Srikantia, and Gopalan (1968) measured the IQs of 19 severely malnourished Indian children several years after they had recovered from malnutrition. The 19 children ranged in age from 8 to 11 years at the time of measurement. There were marked differences between malnourished children and a larger control sample on both IQ tests and tests of intersensory integration. Poorer performance was most apparent for younger malnourished children. The largest differences were found on tests of perceptual and abstract abilities.

Birch, Pineiro, Alcalde, Toca, and Cravioto (1971) reported on intellectual differences between 37 Mexican children with severe malnutrition and 37 siblings who had never experienced malnutrition. WISC IQ scores were obtained on the malnourished children and on the sibling controls at least three years after the malnourished boys had left the hospital. The authors reported that malnourished boys had significantly lower IQs than their siblings.

Hertzig et al. studied both the intellectual development and the school functioning, including social functioning, of 71 Jamaican children who had suffered from malnutrition during infancy (Hertzig, Birch, Richardson, & Tizard, 1972; Richardson, 1974; Richardson et al., 1972; Richardson, Birch, & Ragbeer, 1975). There were two control groups: a sibling control group and a matched school age control. The WISC IQ test was used to measure cognitive abilities, and behavioral assessments were based on parent and teacher evaluations of the children. The investigators found that malnourished children performed more poorly than the matched controls, including the sibling controls, on the WISC full-scale and performance IQ tests. The authors noted, however, the important differences in the environments of the malnourished and nonmalnourished children (see the section above). On behavioral assessments, malnourished children were rated as less mature, more labile in activity, and less social than the matched comparison children. There were no differences between the malnourished children and their siblings on the behavioral measures.

Galler et al. studied 129 boys and girls in Barbados who had been hospitalized before the age of 1 year with second- or third-degree malnutrition. They were compared with 129 matched controls, schoolmates of the same gender, age, and handedness, at 5–11 years of age. The index (malnourished) children showed a higher rate of school failure than the control children. They scored significantly more poorly on a standardized intelligence test and showed memory deficits.

McLaren et al. (1973) studied the effects of nutritional marasmus on Lebanese village children. Subjects were 30 children who had suffered from acute, severe nutritional maramus, with weights at 53 percent of median standard weight for age. They were hospitalized between the ages of 2 and 16 months. A group of children who had been moderately undernourished for at least 3 years also was included. Control groups consisted of siblings of the malnourished group and a group of unrelated, healthy children matched for age and social-economic conditions. The authors found that at ages 3 to 5 years, the undernourished groups consistently scored below siblings on the Stanford-Binet IQ. The largest deficit was in a group of children who had suffered from severe marasmus, with a mean IQ of 78. The mean IQ of the healthy control group was 99.

One controlled study did not show any significant differences between previously malnourished children and sibling controls on later measured intelligence. Hanson et al. (1971) studied 40 lower social class, South African children with documented histories of kwashiorkor. All had been hospitalized between the ages of 12 months and 4 years. At approximately age 11, the two groups were compared on the New South African Individual Intelligence scale. Both groups scored significantly below the normal level, with means of 77 for the index children and 78 for siblings. Both groups were judged as chronically undernourished (Lloyd-Still, 1976).

One study which did not employ a control group nonetheless documented dramatic effects of malnutrition and environmental deprivation on cognitive development. Monckeberg (1968) studied 14 Chilean children who had suffered from severe marasmic malnutrition between 3 and 11 months of age. They were followed up at age 3 to 6 years on the Binet intelligence scale. The mean IQ was 62. No child scored above 76.

4.2.2.4 Effects on social development There have been few studies on the effects of malnutrition on children's social behavior. Several recent studies have considered effects of chronic malnutrition and will be considered in a later section. Two retrospective studies examined the long-term effects of acute, severe undernutrition in infancy for later social development. These were the studies by Richardson and colleagues in Jamaica (1972), and by Galler et al. in Barbados (1983b). In addition, Stoch et al. (1982) conducted a long-term follow-up of the malnourished children they had followed since infancy (see Stoch & Smythe, 1963).

The design of the study by Richardson et al. was described earlier (see Hertzig et al., 1972), in connection with the summary of research on severe malnutrition and intellectual development. In the study, 74 children who had been severely malnourished during infancy were compared with siblings and matched non-sibling controls in later childhood. A total of 58 index children, 58 matched nonsibling controls, and 31 siblings were available for a study of the effects of the early malnutrition on school peformance and behavior. The subjects' social behavior at school was assessed at ages 5 to 11 years. Behaviors related to classwork and general conduct were studied. Peer evaluations focused on social involvement with and acceptance by others.

Children who had been hospitalized for severe malnutrition in early life showed an identifiable constellation of behavior problems at school age. The children were characterized by poorer attention and memory, greater distractability, and less classroom involvement than control children. They were rated as less popular by their peers and less cooperative by their teachers. They showed poorer academic performance and more conduct problems than the control children. Most often, their behavior was described as passive, quiet, and withdrawn (Richardson et al., 1972, p. 283).

The study by Galler et al. (1983b) was based on the sample described in the previous section. A group of 129 Barbadian children, ages 5 to 11 years, who had suffered from moderate to severe protein-energy malnutrition in infancy, were compared with 129 matched control children with respect to school behavior. Data were gathered from questionnaires administered to teachers. Questions concerned classroom behavior, social interaction, health, and emotional maturity. Previously malnourished children showed significantly poorer attention, social interaction, and general health (poorer appearance, sleep, and current nutrition) than control children. The behavioral deficits associated with malnutrition were greater in boys than in girls, and unrelated to differences in intellectual ability.

The recent study by Stoch et al. (1982) also shows long-term behavior impairment resulting from early, severe malnutrition. We discuss this study in Chapter Five (Section 1.0).

The above studies, particularly when considered in conjunction with the research on infant behavior and development, are very important in providing a base for the present research. They strongly suggest that early malnutrition has long-term and adverse consequences for social and emotional development, with attention, social responsiveness, and emotional adjustment significantly affected.

4.3 Malnutrition Secondary to Neonatal Illness (Comparative, Prospective Design)

4.3.1 Design characteristics

Several studies have involved identification of children who experienced a neonatal illness that interfered drastically with nutrient intake and/or retention in the first months of life and who, as a result, could be classified as malnourished secondary to neonatal illness. Such studies also have generally been comparative, nonexperimental, and prospective. Although data collection generally has been retrospective, from childhood to infancy, data analysis has been prospective with the selection variable being presence or absence of early undernutrition, and the outcome variable developmental impairment.

These studies, like the retrospective studies of severe malnutrition, have included a variety of methods for constructing equivalent index and control groups, generally involving hospitalized nonmalnourished controls (e.g., Beardslee et al., 1982). They are vulnerable to the same problems of internal validity as are the first class of studies. They attempt, however, to reduce the scope of selection bias by ruling out the possibility that experimental and control subjects differ greatly in experiences; specifically, by ensuring that control subjects have experienced similar degrees of hospitalization. Most importantly, they refine the construct malnutrition by removing it from the social-environmental context of poverty and low social status. They help to test the internal validity of associations between malnutrition and developmental outcome by addressing the question of how early and severe malnutrition will relate to later functioning when it occurs outside of the usual social-environmental context. In so doing, however, they prevent the generalizability of the results to endemically malnourished populations.

4.3.2 Review of research

In general, findings on cognitive development have been negative. Valman (1974) compared eight children who had had neonatal resection of the ileum with eight sibling controls and 13 children with cystic fibrosis on the Draw-a-man intelligence test. Age at time of testing ranged from 3 to 14 years. There were no differences between the groups in average Draw-a-man scores. Lloyd-Still (Lloyd-Still et al., 1974) reported results of a similar study of children with cystic fibrosis and other neonatal intestinal disorders. Among children tested between 2 and 5 years of age, previously malnourished children scored lower on the Merrill-Palmer test of intelligence than did sibling controls. The most marked abnormalities were in fine motor function. Above the age of 5 years, there were no significant differences between the groups.

Ellis and Hill (1975) studied two groups of children with cystic fibrosis. One group had been malnourished in the first year, while the other group had not been malnourished. The subjects were tested at ages 7 to 10 years. There was no significant difference between the groups in full-scale WISC IQ. Means were 104.1 for control subjects and 99.3 for experimental subjects. There was a significant difference on the backward coding subtest, with malnourished children scoring lower.

Beardslee et al. (1982) carried out a follow-up study of the Lloyd-Still et al. subjects. A group of 31 children and adults who had been severely malnourished in infancy due to intestinal disease were compared to sibling controls. The sample included previously malnourished children who had been under 5 years at the time of the first assessment and had shown significant cognitive deficits, and children who had been over 5 at the time of the first assessment and had not shown cognitive deficits.

Follow-up assessments were on psychometric intelligence, academic performance (judged by teacher ratings), and developmental history, based on parent questionnaires. There were no differences between index and control children on any measures.

One study documented long-term effects of early undernutrition due to organic insult. Klein, Forbes, and Nader (1975) studied the learning abilities and social adjustment of 50 children between the ages of 5 and 14 years who had had pyloric stenosis as infants.

Index children were compared to 44 siblings and 50 matched controls. Learning abilities and adjustment were affected. Previously malnourished children showed poorer short-term memory and attention than nonmalnourished controls, and performed more poorly on the Vineland scale of social maturity.

While the studies suggest that early malnutrition occurring in the context of a nondeprived environment does not necessarily have long-term effects on child development, there are two important points to be made about this research. These concern the sensitivity of the dependent variables, and the nature of the subjects' environments.

First, in these studies, there was little attempt to develop a method of assessment which would be most sensitive to the attentional, motivational, and emotional characteristics that have been shown to be strongly associated with undernutrition. The one study which specifically tested for attentional deficits (Klein et al., 1975) did identify differences between malnourished and control children, and the results of the Ellis et al. study also suggested attentional impairments. Second, the studies in which no long-term effects of malnutrition were found cannot be generalized to environments in which there is no rehabilitation. Beardslee et al. interpret their findings as showing the "compensatory role of a supportive environment in reversing the behavioral and intellectual deficits of malnutrition" (p. 1440). The studies show not that malnutrition as it normally occurs has no long-term adverse consequences, but that when malnutrition is treated by nutritional and social support, long-term effects may be moderated.

4.4 Effects of Nutritional Supplementation on Children's Behavior (Experimental Design Involving Random Assignment to Treatments; Quasiexperimental Design Involving Nonrandom Assignment to Treatments; Longitudinal Sequential Designs Using Multiple Cohorts)

4.4.1 *Design characteristics*

In recent years there have been a large number of studies in which nutritional status has been experimentally manipulated, and the effects of nutritional status on behavior examined. Several studies have used true experimental designs in that, following sample selection, subjects have been randomly assigned to treatments. Others have been quasiexperimental. Here, subjects are assigned to treatments but not individually on a random basis. We noted in a previous section that in the INCAP Guatemala study, villages were assigned to treatment conditions so that once a village had been assigned, assignments of persons within the villages were completely determined. In quasiexperimental designs, the unit of assignment (e.g., the village or school) is not the same as the unit of analysis (i.e., the individual).

Virtually all of the animal studies reviewed in this chapter used experimental designs. In general, in animal studies the nutritional manipulation involves a nutritional deprivation, prenatal and/or postnatal. This often is followed by a reversal component, that is, institution of food supplementation. In some studies, interactive effects of nutritional deprivation/supplementation and environmental stimulation/deprivation are considered. A major methodological concern in the animal studies is (a)external validity and the relationship of humanly imposed malnutrition to naturally occurring malnutrition, and (b)the construct validity of associations between malnutrition and behavior, due to the fact that most nutritional manipulations involve introduction of or removal of extranutritional stimulation. These issues were discussed earlier in this chapter in the section on methodology in animal malnutrition research.

Human studies that have used experimental designs include the Bogota study (Herrera et al., 1980), the Louisiana WIC study (Hicks et al., 1983), the Bacon-Chow study (Joos & Pollitt, 1983), and the New York study (Rush et al., 1980). In each of these studies (reviewed below), individual mothers or children were assigned to experimental or control groups and followed through the child's infancy. The major concerns about nutritional supplementation studies have been described (Barrett, 1984b; Barrett, 1984c). These include two threats to the construct validity of the research: (a)the problem of inadequate preoperational conceptualization of the independent variable (nutritional supplementation), and (b)failure to distinguish the construct from the level of the construct. In addition, there are the possibilities for the threats to internal validity resulting from the control group's recognition and/or the investigator's recognition that the control group is receiving an inferior treatment. These problems were described earlier in Chapter Two as diffusion or imitation of treatment, compensatory equalization, compensatory rivalry, and resentful demoralization. Finally, there is the particular threat to statistical conclusion validity of hetereogeneity of scores on a dependent variable due to individual differences in application of treatment. This occurs when the investigator does not have complete control over the administration of treatment.

A third concern is limited generalizability. We have touched on this issue in discussing the strengths and weaknesses of experimental versus nonexperimental designs. The results of intervention studies tell us what could happen in an endemically malnourished population were the level of feeding to be raised, but they do not tell us what a deficit of those nutrients means for the child's behavior. Including a missing nutrient in a diet cannot be assumed to have the opposite effect of taking away the nutrient. Further, because of random assignment, the environmental conditions that covary with but also may make possible the adverse effects of malnutrition (i.e., which provide the setting conditions for the adverse effects) may not be identified. In the absence of supplementation effects, real (although complex) relationships may be obscured.

Despite these concerns, supplementation studies are widely considered the most reliable and, in particular, the most internally valid means for assessing empirical relationships betwen nutritional and behavioral variables (see Rush, 1984).

4.3.2 Review of research

There have been seven major studies on the effects of nutritional supplementation on the behavior of children at risk for undernutrition. These are the Guatemala INCAP study; the Bogota, Columbia, study; the Tezonteopan, Mexico, study; the Cali, Colombia, study; the New York study; the Louisiana WIC study; and the Bacon-Chow Taiwan study. All have demonstrated significant effects of protein-energy supplementation on behavioral development.

The Guatemala study (Klein, 1979; Klein et al., 1974, 1977)

examined the impact on child development of nutritional supplementation during pregnancy, lactation, and the first several years of the child's life. The original intent was to compare the effectiveness of a protein-rich supplement and a protein-free (calorie-only) supplement. Children and mothers in two villages in eastern Guatemala were assigned to a protein-calorie supplementation treatment, while children and mothers in two other villages were assigned to the calorie-only treatment. Villages were matched on a number of demographic variables.

Supplementation began for pregnant women at the beginning of pregnancy and continued through lactation, and for children began immediately after birth. Supplements were administered at central food centers, ad libitum.

Analyses (Klein, 1979) showed that supplemental calories predicted level of cognitive development at preschool age. Subjects were divided into high, middle, and low levels of supplement ingestion, based on the amount of supplemental calories during the mother's pregnancy and lactation, and during the child's early life. By 4 years of age, there were strong associations between energy intake and cognitive function. Better-supplemented children showed better vocabularies, discrimination learning, and overall cognitive test scores. Further analyses showed that the effects of supplementation on test performance were strongest for children of the poorer social-economic circumstances.

It is this data base which provided nutritional data for the present study, reported in Chapter Four, on nutritional supplementation and social and emotional development at school age.

The Bogota study (Herrera et al., 1980; Waber et al., 1981) involved mothers and children in the barrios of Bogota, Colombia. There were two types of intervention: a nutritional intervention and an educational intervention. Subjects were assigned at random to one of several conditions, differing in type of intervention (supplementation, education, both, or neither) and timing of intervention (beginning at 6 months of pregnancy, beginning at the child's birth, or beginning at age 6 months). Supplements rich in protein, calcium, vitamins, and iron were distributed to families weekly.

Significant effects of supplementation were discernible by 15 days. In comparison to supplemented children, nonsupplemented infants showed heightened reactivity to aversive stimuli. By 4 to 8 months of age, nonsupplemented infants showed more apathetic behavior. At 3 years, children who had received food supplementation had higher scores on locomotor, personal-social, speech and language, eye-hand coordination, and performance tests on a standard developmental assessment, and also had a higher general quotient. Duration of supplementation was important. Subjects who received food supplementation only up to age 6 months performed no better by age 3 than nonsupplemented controls.

The Tezonteopan study (Chavez et al., 1975; Chavez and Martinez, 1979) involved two groups of mothers and infants. A group of 20 mothers who had given birth to normal infants, had 1–4 previous children, and fell into a specified range for height and weight were selected for the nonsupplemented, control group. A total of 20 women matched on relevant variables were chosen one year later, as they were beginning their pregnancies, for the supplemented, experimental group. Supplemented women received daily protein-calorie supplements beginning at the sixth week of pregnancy and continuing through lactation. Infants were supplemented with whole milk and various prepared baby foods beginning at age 3 months.

By 8 months of age, supplemented children were more active than nonsupplemented children. By age 1 year, they spent less time crying and clinging, and at 18 months were significantly more independent. They were carried less, spent more time out of doors, and played more. Supplemented children also were more demanding, and more successful at eliciting both maternal and paternal responses. These qualities appear to have resulted in their being fed more often.

The Cali, Colombia study (McKay et al., 1978) differed from the other supplementation studies in that the nutritional intervention was linked with a "psychoeducational" intervention in all treatment groups. Thus, the study does not allow one to draw inferences about the role of improved nutrition alone on later behavior. The study does, however, provide some information about the importance of the timing and duration of intervention for high-risk children, and this issue is addressed in Chapter Five of this book.

The New York City study (Rush et al., 1980) involved the

administration of three different types of treatments to mothers at risk for delivering low birth weight babies. Criteria for sample selection were low prepregnancy weight, previous low birth weight infant, or previous day protein intake below 50 grams. Mothers were assigned to a high-protein supplement group, a low-protein complement group, or a control group. The supplement contained 470 calories and 40 grams of protein while the complement contained 320 calories and 6 grams of protein. Results of the intervention for child growth were complex, depending on whether or not the child was premature, and on the timing of the supplementation, as well as on the type of beverage (high vs. low protein). Effects on behavior were more readily interpretable: High protein-calorie supplementation was associated with better performance at 1 year of age on measures of attention, persistence, and visual habituation.

A recent supplementation study conducted in Louisiana (Hicks et al., 1982) examined the effects of participation in the WIC supplementation program on children's behavior at school age. Subjects were sibling pairs in which one sibling participated in the WIC program for at least four years (on average) beginning prenatally, and the other participated for at least three years, beginning after age 1. The prenatally supplemented group showed higher IQs, better attention span, better visual-motor integration, better school performance, and fewer behavior problems when measured between 6 and 11 years of age than the later supplemented group.

A final study is the Bacon-Chow study of maternal supplementation (Joos & Pollitt, 1983), which was carried out in Taiwan. Women were assigned at random to two treatment groups. One received a high-protein, high-calorie supplement (800 calories, 40 grams of protein per day) and the other was offered a placebo. All women in the study were receiving low levels of calories and protein prior to the study. Infants were assessed at approximately 8 months of age for mental and motor development. There were no differences in mental scores for the two groups of infants. Infants of supplemented mothers, however, had higher motor scores.

The supplementation studies provide strong support for the hypothesis that chronic as well as acute malnutrition can have long-term effects on the child's development. In each of the studies cited, children were from populations known to be receiving inadequate food energy. In each study, the provision of additional food energy resulted in changes in behavior. In studies that focused on infants, major changes were in neuromotor maturation and coordination. Studies that followed up children through the early school years identified effects on performance on tests of intellectual development.

The processes whereby early undernutrition can affect the child's behavior are complex and include both the direct effects of insufficient food energy on the child's physical condition and capacities, and also the indirect effects of reduced food intake, mediated by changes in interaction with the environment and increased susceptibility to illness. The issue of mediating processes is discussed further in Chapter Five.

4.5 Studies of Infants Who Are Failing to Thrive

Only recently have investigators recognized that the extensive literature from the United Status, United Kingdom, and Australia on children with so-called nonorganic failure to thrive (NOFT) is homologous with that on children from the developing world with diagnosed protein-energy malnutrition (PEM). Although the parallels between these two bodies of data are striking, there are, for historical reasons, systematic biases in the failure-to-thrive literature that suggest the need for caution in interpreting the reported developmental and behavioral findings.

Failure to thrive is not a diagnosis but a symptom. The term is used to describe children whose growth deviates from the expected norms for age and sex. Traditionally, such growth failure has been dichotomized into "organic" failure to thrive, where the child's impaired growth is ascribed to a serious acute illness or chronic major organ system dysfunction (such as gastrointestinal or renal disease) and nonorganic failure to thrive (NOFT), where no primary medical condition sufficient to account for the growth failure can be identified. Until the late 1960s such NOFT was ascribed to inadequate affective nurturance in either institutions or homes (Leonard, et al., 1966; Patton & Gardner, 1962; Spitz,

1945), a syndrome variously described as "hospitalism" or "maternal deprivation." Inadequate emotional stimulation was thought to produce growth failure through endocrine pathways, independent of nutritional status (Powell, Brasel, & Blizzard, 1967). The pathogenic role of inadequate dietary intake in producing growth failure in NOFT was recognized only in 1969 (Whitten, Pettit, & Fischoff, 1969). However, it was tacitly assumed that in industrialized countries such nutritional deprivation could only occur as a function of idiosyncratic parental psychopathology, producing a disturbed parent-child interaction that resulted in the child not receiving adequate food. Only recently have investigators recognized that children labelled as NOFT represent the extreme end of a spectrum of nutritional growth failures that afflict almost 10% of low-income children under age 5 in the United States, according to several regional studies (Massachusetts Nutritional Survey, 1983; Trowbridge, 1984). Concurrently, researchers have noted that there is no pathognomonic parental psychopathology associated with NOFT, but that the predisposing social conditions are the same as those predisposing the occurrence of PEM in the third world—poverty, social isolation, large household size, recurrent infections, and low birthweight (Bithoney & Rathbun, 1983). However, because of the a priori assumption that psychosocial insults were primary in producing nonorganic failure to thrive, much of the early literature on behavioral and developmental concomitants and sequelae of NOFT may be biased. This issue is addressed further below.

4.5.1 Design characteristics

Failure to thrive (NOFT) studies represent the most heterogeneous group of studies with regard to research design and thus a general description of the design category in which they fall is not attempted. Studies have been conducted on the relationships between NOFT and a number of variables related to child behavior: maternal feeding behavior with the child, child temperament and behavior characteristics, mother-child interaction (i.e., in a nonfeeding context), and child cognitive development. Studies on these topics have included comparative studies of NOFT and non-NOFT children at the time of diagnosis (Pollitt & Eichler,

1976), retrospective studies examining maternal reports about the child prior to onset of NOFT (Kotelchuck, 1980), longitudinal, prospective studies comparing NOFT with non-NOFT children (Fitch et al., 1976; Glaser et al., 1968), and nonexperimental, non-control group reports of characteristics of infants that are currently failing to thrive (Powell and Low, 1983).

A major methodological concern in much of the NOFT research is the tautological nature of inquiries into NOFT. While recent investigations have selected children for study of NOFT on anthropometric criteria alone (Drotar et al., 1985), early researchers required developmental delay and/or behavioral disturbance or obvious family psychopathology as criteria for diagnosis of NOFT (Whitten et al., 1969). Malnourished children without obvious developmental/behavioral or family derangements at the time of presentation may have been excluded from clinical samples of NOFT, thus biasing follow-up findings in the direction of excess morbidity.

There are other methodologic difficulties in using reports of research on failure to thrive to elucidate the effects of malnutrition on behavior, although many of these difficulties are similar to those found in evaluating studies from the developing world. A first problem and an extremely important one affecting statistical conclusion validity is that there are no standard anthropometric criteria across studies for classifying children as failure to thrive (whether organic or nonorganic). Definitions used have ranged from depressed rate of weight gain for 10 or more days (Vietze et al., 1980) to height and weight below the third percentile (Elmer, 1960). Thus the severity of malnutrition experienced by children diagnosed as NOFT varies widely from study to study. This means that failure to replicate findings on NOFT and behavior may reflect differences in the criteria used to diagnose NOFT. In addition, there are three particular threats to the internal validity of NOFT studies, all involving the problem of selection effects. First, studies such as Rosenn et al. (1980) and Singer and Fagan (1984), which compare children with nonorganic failure to thrive to those with organic failure to thrive, may be confounded by the fact that children in the organic category may also be suffering from malnutrition secondary to gastrointestinal disease, which afflicts 8–19% of reported series of children hospitalized for failure to thrive (Berwick, 1980), so that in such studies possible nutritional effects on outcome may be obscured.

Second, studies of NOFT vary in whether children with low birthweight are excluded or included in the clinical sample. As in studies from the developing world, there is possible confounding of the effects of postnatal malnutrition with effects of the complex of prenatal conditions leading to prematurity, and with the effects of the myriad non-nutritional perinatal medical complications (such as asphyxia and intraventricular hemorrhage), which may affect the growth and development of low birthweight infants, independent of later nutrition.

Finally, the majority of studies of failure to thrive draw on samples of children hospitalized in infancy and early childhood, which may confound the adverse results of malnutrition with those of hospitalization in early life (Quinton & Rutter, 1976).

4.5.2 Review of research

The short- and long-term cognitive and behavioral concomitants of nonorganic failure to thrive are strikingly similar to those associated with diagnosed childhood protein-energy malnutrition in the developing world. Children with failure to thrive are noted to demonstrate aberrant social interaction, and to be apathetic, irritable, and inattentive (Pollitt, 1975; Powell & Low, 1983; Rosenn, Loeb, & Jura, 1980) at presentation. When re-evaluated months to years after initial diagnosis, these children show, on average, depressed scores on psychometric assessments of cognitive development (Chase & Martin, 1970; Elmer Gregg & Ellison, 1969; Fitch et al., 1976; Glaser et al., 1968; Hufton & Oates, 1970; Singer & Fagan, 1984), with somewhat better outcomes when families receive psychosocial supportive interventions (Drotar et al., 1985; Field, 1984), or when NOFT children are compared to controls from equally disadvantaged families (Mitchell et al., 1980). In follow-up studies of NOFT, behavioral or psychological disturbances have been found in 28% (Glaser et al., 1968) to 48% of children (Elmer et al., 1969; Hufton & Oates, 1977; Oates, Peacock, & Forrest, 1985). Two investigators also noted academic delays (Glaser et al., 1968; Hufton & Oates, 1977).

In summary, the literature on nonorganic failure to thrive

suggests that in the developed and in the developing worlds, children who suffer early nutrition-related growth failure in the context of social disadvantage are at increased risk for later deficits in social-emotional functioning, cognitive abilities, and academic achievement.

4.6 *Studies on Intrauterine Growth-retarded (IUGR) Infants (Comparative, Prospective Designs; Longitudinal Prospective Designs)*

As discussed in Chapter One, only asymmetric intrauterine growth retardation, manifested by depressed ponderal index, can be regarded as a relatively pure model of intrauterine malnutrition, since the other forms of low birthweight, including those resulting from prematurity and from symmetric intrauterine growth retardation, reflect so many possible biologic derangements in the maternal-fetal unit that the effect of nutrition is difficult to isolate in a particular case. Even so, the literature on asymmetric intrauterine growth retardation must be interpreted with some caution when ascribing observed behavioral or cognitive sequelae to intrauterine undernutrition.

4.6.1 Design characteristics

In studies comparing asymmetric IUGR with non-IUGR term infants, there are two major areas of difficulty. First, infants with asymmetric growth retardation are at increased risk for perinatal metabolic disturbances such as hypoglycemia and polycythemia, disturbances which themselves may inflict damage on the central nervous system and thus on later behavior independent of prenatal nutritional experience (Gaston, 1982). This is a significant threat to the internal validity of IUGR studies. Second, when behavioral measurements are obtained several years after birth, it is not possible to determine whether observed differences between asymmetrically growth-retarded infants and appropriate-for-gestational-age controls should be ascribed to prenatal nutritional experience alone or, in addition, to nutritional and social experiences in the years intervening between birth and assessment. This is a particularly troublesome problem since asymmetrically growth-retarded infants may be poor eliciters of parental nurturance, and thus receive less adequate feeding and social stimulation during postnatal life than their more attractive appropriate-for-gestational-age peers (Als et al., 1976; Pollitt, 1975). This second problem can be conceived of as a threat to the construct validity of putative causes and effects. That is, it *is* prenatal nutritional status that is responsible for differences between index and comparison groups in later outcome; however, the pathways by which effects occur may be different from those conceptualized by the researcher.

4.6.2 Review of research

Studies reviewed below include those in which index children were asymmetric growth retarded. However, in one study (Brazelton, 1977) low birthweight alone was used to identify subjects as IUGR; in this study, both symmetric and asymmetric IUGR are represented.

4.6.2.1 Behavioral characteristics of underweight newborns Studies in several cultures have identified behavioral problems in newborns who were born at term but were underweight. The behavior problems identified indicate a consistent pattern of adverse outcomes. Als et al. (1976), studying a U.S. population, compared infants of 10 days or less with low ponderal indices (PI), a standard measure of weight-for-length defined previously, to infants with normal ponderal indices. While birthweight was not controlled, the low PI group was only moderately low birth weight (average weight = 2668 grams). Thus, they were identified as intrauterine growth retarded (IUGR) on the basis of low weight for length. On the Brazelton Neonatal Behavioral Assessment Scale, compared to normal PI babies, the IUGR children showed significantly suboptimal coordination, motor activity, and muscle tone. They were less socially responsive, less active, and less consolable than normal-weight infants. Brazelton et al. (1977) examined behavioral characteristics of nutritionally deprived Guatemalan infants who were full-term but low birth weight. The children were studied at one month of age. The children had poor muscle tone, showed little vigor of movement when in higher states of arousal, and had difficulty orienting to visual stimuli and eliciting social response. The findings of poor attention, motor development, and social responsiveness are highly consistent with those of Als et al.

A third study showing striking behavioral differences between underweight and normal-weight infants was carried out in Puerto Rico by Lester and colleagues (Lester et al., 1986). In the study, infants identified as IUGR on the basis of low birth weight and/or ponderal index were compared with normal-weight infants. All testing on the NBAS was between 18 and 60 hours of life. IUGR infants were less responsive, less able to process visual and auditory stimuli, and motorically immature, compared to normal-weight infants.

A larger sample study on the effects of IUGR is from Villar and colleagues (1984). Postnatal growth and development were compared in three groups of infants: one group with normal birthweight and normal ponderal index, one group of low birthweight and low ponderal index (IUGR-LPI), and one of low birthweight but normal ponderal index (IUGR-API). Of particular interest for the present discussion is the comparison between the low ponderal index, low birthweight and the normal ponderal index, low birthweight groups, since the low ponderal index group represents a group that can be unequivocally identified as prenatally undernourished, in contrast to the normal ponderal index, low birthweight group, who may have suffered from multiple insults in addition to chronic fetal undernutrition.

Subjects were 205 term infants enrolled in the INCAP longitudinal study in Guatemala, discussed in the section on nutritional supplementation. A total of 146 children were normal birthweight. Of the 59 low birthweight children, 21 were IUGR-LPI and 38 were IUGR-API. While at 3 years of age the IUGR-API group scored the lowest on seven of eight measures of behavioral development, the IUGR-LPI group scored lower than the normal group on all measures. Thus, there was a consistent pattern of outcome with normal birthweight children scoring highest, IUGR-API scoring lowest, and IUGR-LPI scoring at levels between the two.

The IUGR-API group represents a group that is symmetric growth retarded. This group, stunted in length and underweight for gestational age, is at the greatest developmental risk. The IUGR-LPI group represents a group that is asymmetically growth retarded. They are wasted but not stunted. This group showed long-term effects of prenatal undernutrition, although the effects on behavior appear to be not as severe as in proportional IUGR. Villar et al. also showed that while the IUGR-API group remain shorter, lighter, and have smaller head circumferences than normal children by age 3, the IUGR-LPI group shows a catch-up in weight to nearly normal values.

4.6.2.2 Mother-infant interaction There are few data on the effects of low birth weight on mother-infant interaction. Inferences about the effects of prenatal undernutrition on the caregiver generally are made on the basis of data from studies which have examined the newborn in interaction with an experimenter . The most recent evidence comes from the research by Garcia-Coll and colleagues in Puerto Rico. In their study, effects of IUGR infant status on mother-child interaction were found for teenage mothers. Teenage mothers with IUGR infants were more likely to have feeding problems with their infants than were teenage mothers with non-IUGR infants. Mothers reported that the infants spit up and did not ingest a high amount of formula, even for prolonged feedings. In addition, they reported that they had difficulty understanding the infant's feeding cues (Garcia-Coll et al., 1984).

A most interesting and recent source of information about interactive problems in undernourished children is a set of studies which focuses on the undernourished child's cry and its effects on the caregiver. Lester (1979) has examined the relationship between newborn nutritional status and cry characteristics in full-term children who were identified as IUGR on the basis of the ponderal index. Children were measured both on the Brazelton scale and on acoustic features of the cry. As was the case in earlier studies, the undernourished children showed underdeveloped motoric processes, state organization, and physiologial organization (for example, startles, tremors, and skin color changes) compared to adequately nourished children. In addition, there were clear differences in cry characteristics. When elicited cries were examined, undernourished infants showed shorter initial cries followed by a longer latency to second cry (expiratory period). There were significant differences in fundamental frequencies and harmonics as well, with the undernourished infants having higher fundamental frequencies and fewer harmonics.

The significance of such results is suggested by a study of the effects of the infant's cries on the caregiver. A total of 30 men and 30 women listened to tape-recorded cries of eight infants who were identified as high risk or low risk, based on prenatal and perinatal factors (maternal, placental, and fetal) (Zeskind & Lester, 1978). The adults were asked to rate the cries on different dimensions. The cries of the high-risk infants were rated as significantly more distressing, arousing, urgent, grating, sick, piercing, discomforting, and aversive than the cries of the low-risk infants. Of particular relevance is the fact that, like the cries of malnourished infants, the cries of the high-risk infants were characterized by shorter duration, longer latency, and higher fundamental frequencies. The results suggest that one of the effects of prenatal malnutrition, as well as other perinatal factors, is to inhibit the development of behaviors that elicit positive emotional responses on the part of the caregiver and, thus, perhaps, to impede the development of the parent-child relationship.

4.7 Correlational Studies of Chronically Malnourished Children (Naturalistic, Cross-sectional Studies)

4.7.1 Design characteristics

Naturalistic, cross-sectional studies involve identifying a target population, obtaining a representative sample from that population, and then, within that group, examining relationships between the independent and dependent variables of interest. In malnutrition research, populations assumed to be chronically malnourished and/or at risk for malnutrition have been identified, and then within the population, associations between nutritional history and one or more measures of behavioral development have been studied. Examples are Cravioto et al.'s study (1966) of the correlates of height in Guatemalan children, and Freeman et al.'s (1977) study in Guatemala, using cross-sectional anthropometric data of the relationships between height and head circumference and cognitive development in children from the INCAP sample.

Cross-sectional studies are vulnerable to many of the same threats to internal validity as are retrospective comparative studies. In addition, they are at risk for poor statistical conclusion validity, since they are often less powerful than studies in which one begins by identifying children as abnormal based on either the independent or dependent variable (see Section 4.6 in Chapter Two). Still, cross-sectional, naturalistic studies are a logical way to begin to explore the complex relationships among variables resulting from, correlating with, and even predicting malnutrition. Such studies often are the first step in beginning to test hypotheses about the effects of improved or inferior nutrition in specific populations.

4.7.2 Review of research

Several studies have examined correlations between measures of long-term undernutrition and child cognitive development. Two studies were carried out in Guatemala. Cravioto et al. (1966) studied 143 village children, ordering them by height. The upper- and lower-quartile groups then were compared on a number of cross-modal, sensory-integration tasks. The rationale for comparing the upper and lower quartiles was that the upper-quartile children presumably represented the group with the least likelihood of having experienced a significant degree of malnutrition earlier in life, while children in the lower quartile were assumed to have the greatest likelihood of having been at earlier nutritional risk. Similar comparisons were made in an urban, upper-middle class population, where there was assumed to be no association between height and previous nutritional history (i.e., where height differences were assumed to reflect primarily quantitative, genetic factors). Thus, if upper- versus lower-quartile differences were found in the rural population but not in the urban population, results could be interpreted in terms of the effects of malnutrition on the criterion behaviors.

The authors found that for children in the rural villages, subjects in the lower quartile for height showed poorer intersensory integration—visual-haptic, visual-kinesthetic, and haptic-kinesthetic—than children in the highest quartile. There were no differences in the upper-class group. Results were interpreted as supporting the hypothesis that malnutrition results in a lag in the development of sensory integrative capacities.

A second Guatemala data collection was performed by Klein et al. (1974). Height and head circumference were correlated with mental and motor scores on standardized infant developmental tests for children from villages in eastern Guatemala. The study was conducted as part of the INCAP longitudinal study. Height of the child at ages 5 and 7 was correlated with vocabulary at ages 5 and 7, as was head circumference. Relationships between height and head circumference and other cognitive tests were inconsistent, depending on sex and age of testing.

Freeman et al. (1977) examined the associations between height, head circumference, and measures of language, memory, and perception for 1083 children from the INCAP study, ages 3 and 4 years. In general, both height and head circumference were predictive of the cognitive measures. Prediction generally continued to be significant even with a measure of social-economic circumstances controlled.

In a study in Nigeria, Ashem and Janes (1981) compared 45 well-nourished urban children with 28 reasonably nourished urban and rural children and 55 poorly nourished urban and rural children ages 2 to 6 years. Nutritional status was based on height and weight for age. The well-nourished group scored significantly higher than the reasonably nourished group on the McCarthy Scales of Abilities, and the reasonably nourished scored higher than the malnourished.

A somewhat different approach was taken by Hoorweg in a study of Ugandan children (1976). Subjects were three groups of 20 children admitted for protein-energy malnutrition to the Kampala clinic, and 20 comparison children who had not suffered from malnutrition. Children in the malnourished groups had been admitted to the clinic at ages 8–15 months, 16–21 months, and 22–27 months, respectively. One analysis focused on differences between malnourished and comparison groups on child outcomes at adolescence. A second addressed the question of whether acute or chronic components were most responsible for behavioral deficits in the undernourished. A third addressed the question of the timing of malnutrition as related to later development. Malnourished children scored significantly lower than well-nourished children on a number of standardized intelligence tests and subtests including the Raven matrices, and on the Lincoln Oseretsky motor development scale. To examine acute and chronic components of malnutrition, medical indicators of malnutrition at and just after admission were factor analyzed. Two dimensions emerged. One factor was defined by weight loss after admission, serum protein level, edema, skin changes, and overall clinical severity. The second component was defined by expected weight for age, weight loss after admission, and hemoglobin level. The first component was identified as the acute component and the second as the chronic component. Analyses of relationships between each component and measures of cognitive development showed that scores on the chronic dimension were consistently correlated with intellectual functioning, while scores on the acute component were not. The study demonstrates the importance of chronic undernutrition for intellectual development. Differences in age of admission for hospitalization were not related to differences in cognitive development.

One large study of chronic malnutrition was carried out in the United States (Edwards & Grossman, 1981). The data were collected as part of the U.S. Health Examination Survey, conducted by the National Center for Health Statistics. Subjects were a representative sample of American children, 6 to 11 years of age. Children represented all SES levels and included both black and white children. Children from single-parent families, adopted children, and institutionalized children were excluded. Anthropometric variables were used, along with social-demographic variables and other measures of health to predict intellectual development at school age.

The authors found that child height was a significant predictor of intelligence and achievement test performance. Variations in height were more strongly associated with later outcomes for lower-class children than for middle-class children. Assuming that variation in height in the high-income sample is due primarily to genetic factors, while in the lower-class sample it is due to both genetic and nutritional factors, the difference in coefficients in the two samples suggests a strong nutritional influence on intellectual function.

Finally, although they use a different type of design, the studies by Graves on the effects of moderate malnutrition on infant behavior are considered here because they concern the correlates of mild to moderate undernutrition. Graves conducted a study in West Bengal, India, and a replication in Nepal. In the pilot study (Graves, 1976), 23 children suffering from second-degree malnutrition at the time of testing were examined. For children above the age of 1 year and up to 1-1/2 years, malnutrition was associated with less vigorous exploration and less social interaction with mothers. Maternal responsiveness also was lower to the malnourished children. There were no differences on standard measures of developmental level.

In the replication study (Graves, 1978) in Nepal, Graves again found that 7- to 18-month-old malnourished children showed less exploration than well-nourished controls. They preferred to stay closer to the mother and did not engage in distal contact as frequently as the better-nourished children. In this sample, there were no differences in maternal behavior toward the child.

The studies on the correlates of chronic undernutrition have yielded consistent findings between child growth and cognitive development, but are subject to the potentially strong confounding influences of early illness, between-family environmental differences, and, in the case of older children, differences in school experiences. Nonetheless, the studies, because of their generalizability to current conditions of endemic malnutrition, are an important indicator of the human risks associated with malnutrition in today's world.

4.8 Clinical Intervention Studies (Nonexperimental Designs in Which the Treatment Group Is Its Own Control; Experimental Designs)

4.8.1 Design characteristics

There now have been several short-term intervention studies with severely malnourished children. These have been conducted in Lebanon (Yaktin & McLaren, 1970), Mexico (Cravioto, 1977), Chile (Celedon & De Andraca, 1979; Celedon et al., 1980; Monckeberg, 1979), and Jamaica (Grantham-McGregor et al., 1980). In addition, Grantham-McGregor and colleagues have extended their enrichment program to look at long-term effects of intervention (Grantham-McGregor et al., 1980, 1983).

These studies differ from the supplementation studies not only in that they focus on severe clinical malnutrition, but also in that they assume that the intervention includes a psychosocial component. For example, in Cravioto's study, a psychologist taught each child daily, and nurses were encouraged to play with the children. In Monckeberg's study, the curriculum involved sensory stimulation, affective stimulation, and physical exercise as well as improved nutrition. Grantham-McGregor's study used a cognitive, Piagetian curriculum. Toys were provided to mothers and mothers were encouraged to teach their children through play. Verbal interaction was encouraged.

Thus, neither the experimental nor the nonexperimental clinical intervention studies (i.e., the studies by Celadon and colleagues, in which subjects were their own controls) address precisely the question of the effects of malnutrition on child behavior. Rather, they provide information about the environmental conditions under which nutritional intervention is most likely to positively affect development. In so doing they suggest the important moderating variables in the relationship between malnutrition and behavior.

It is important to distinguish the clinical intervention studies that employ control groups from those that do not. Studies that do not employ control groups but look at change over time in a single group are difficult to interpret because one cannot separate the effects of the intervention (i.e., the improved nutrition) from changes that might take place over time (i.e., maturation and history effects).

4.8.2 Brief review of research

We will review clinical intervention studies in detail at the end of Chapter Four rather than in this section because they are important in elucidating two issues: First, the possibilities for recovery from malnutrition, and second, the importance of the social environment in moderating the effects of malnutrition and/or nutritional intervention. In addition, as indicated above, because all of the controlled studies specifically test for the effects of a psychosocial stimulation component over and above a nutrition component (i.e., nutritional rehabilitation is provided for control as well as experimental subjects), these studies do not bear directly on the question of nutrition-behavior relationships (see Grantham-McGregor, 1984c, for a review).

Briefly, major short-term rehabilitation studies are those by Yatkin and MacLaren in Lebanon (1970), Cravioto in Mexico (1977), Monckeberg in Chile (1979), and Celadon in Chile (1979, 1980). In all of these studies except those of Celadon and co-workers, developmental quotients of malnourished children improved with treatment. Differences were not maintained one year after special care ended in the Lebanon and Mexico studies.

Grantham-McGregor and colleagues have carried out a study on the long-term effects of stimulation and nutritional rehabilitation on Jamaican children (1980, 1983). Children have been studied for five years beginning at age 6 to 24 months. The stimulation program consisted of one hour a day of structured play six days a week while the children were in the hospital. By the end of hospitalization, experimental group children showed significantly higher developmental quotients than nonstimulated, malnourished children. In contrast to children from most short-term studies, subjects in this study continued to show improvements in certain areas of behavioral development—both visual-motor and cognitive—for five years after returning home. However, it is important to note that after the return home, subjects had regular home visits for three years. This continued psychosocial intervention no doubt contributed to the maintenance of the cognitive gains.

The above study involved nutritional intervention in the hospital but not following return to the home. Two years after leaving the hospital, the experimental group subjects were still smaller and lighter than nonmalnourished comparison children. Locomotor development did not show 'catch-up' in the post-hospital intervention period. Other dimensions of behavioral development, including hearing and speech, eye-hand coordination, and reasoning, were comparable to those of well-nourished comparison children.

4.9 Contributions of Different Approaches to the Study of Malnutrition and Behavior

There have been many different kinds of studies addressing the issue of the effects of malnutrition on behavior. The studies have used very different approaches and, as will be seen in the following section, permit inferences about different empirical issues. Together, however, they have provided a powerful base for consideration of the types of behaviors that are likely to be influenced by a history of undernutrition, and the processes which may mediate nutrition-behavior relationships.

4.9.1 Methodological differences and differences in objectives among the different approaches

The different types of studies that have been carried out for examining the effects of malnutrition represent different types of research designs. Studies on the effects of severe malnutrition involving comparisons between previously malnourished children and nonmalnourished children have used nonexperimental, comparative designs, generally with retrospective data collection and prospective analysis. Children initially selected on the basis of poor or adequate nutrition have been compared later in life on behavioral variables of interest. The studies of malnutrition secondary to neonatal illness and the studies of IUGR also have relied primarily on this type of design. Such studies, in addition to having the problem of matching experimental and control groups on potential confounding variables (e.g., differences in social-environmental conditions), are vulnerable to additional confounding nutritional influences, specifically, current nutritional status and micronutrient status, unless such influences are controlled. Nonetheless, these studies, because they help us to explore the consequences of malnutrition in a natural context—that is, without intervention—are important in generating hypotheses about the nutritional factors that influence the development of behavioral competencies. In addition, they indicate the risks associated with malnutrition, considered again as part of the larger complex of conditions under which it may occur.

The cross-sectional studies of chronic malnutrition provide a unique opportunity to examine the correlates of malnutrition in free-living populations and without the influence of selection factors. That is, unlike the comparison studies, there is no initial selection of subjects on the basis of nutritional status. Rather, all subjects are considered, and within the target populations the implications of different degrees of undernutrition are examined.

At this point it is important to reiterate a major distinction between the cross-sectional and the comparative studies. We already have noted that cross-sectional studies enhance the generalizability of research findings because subjects are not selected on the basis of standing on presumed antecedent or consequent variables. In addition, there is a difference in the nature of the relationship between antecedent and consequent variables that is assumed. In studies comparing malnourished and well-nourished children, the categorization of children is qualitative and the hypothesized relationship between independent and dependent variables is nonlinear. That is, children are classified on the basis of presence or absence of a condition (i.e., malnutrition) and compared on a continuous dependent variable. If one conceptualizes the independent variable on a continuum, we may think of the comparison groups as representing extreme positions on the continuum. In contrast, the correlational studies of chronically malnourished populations assume a linear relationship between degree of undernutrition and degree of behavioral impairment on the outcome variable. Absence of a linear relationship weakens obtained correlations. Unless a specific nonlinear model is hypothesized and tested (i.e., by nonlinear trend analysis), real nonlinear relationships may not be detected and important associations thus will be masked.

The failure-to-thrive studies represent a complex set of studies of malnutrition and behavior. The major methodological problem occurs in studies that involve—as part of the initial classification process—documentation of interactional disorders. When mother-child dyads identified in this way are followed up for behavior later in life, analyses are tautological. NOFT studies that use only growth as a classification variable correspond to nonexperimental comparison studies of severely malnourished children. These are prospective in analysis (classification is by standing on a presumed antecedent variable) and may be either retrospective or prospective (i.e., longitudinal) in terms of method of data collection.

It should be noted that nonexperimental longitudinal studies are subject to the same problems of selection (i.e., nonequivalence of groups on other relevant non-nutritional variables) as the studies that are retrospective in terms of data collection, but they are not subject to the historical confounds of the retrospective studies since the investigators keep track over time of the different developments in the experimental and control groups. Thus, differences between groups that develop over time on certain nutrition-related variables such as height, weight, or micronutrient status can be analyzed as outcome variables—that is, possible sequelae of the original classification conditions—and also as variables that may mediate the effects of the classification variable on later behavioral developments.

The nutritional intervention studies for chronically malnourished populations represent a class of studies that, it is generally agreed, have made an important contribution to the studies on malnutrition and behavior. The nutritional supplementation studies have been experimental or quasiexperimental studies, and have provided heretofore unattainable leverage on the question of how the provision of food can affect behavior. Because the studies use or approximate experimental designs, they allow for reasonable inferences of causality concerning nutritional intake and behavioral outcome. The major limitations are as follows.

First, such studies generally involve a combination of nutritional and non-nutritional interventions. Unless such associations are carefully conceptualized—that is, unless it is recognized from the outset that the nutritional intervention actually represents a nutritional/medical/social intervention—inappropriate inferences about the importance of improved nutrition may be drawn. Second, such studies address a hypothetical question, namely, What might be the effects of improving nutrition in this particular population? They do not directly address the question, although they indirectly address it, of how individual differences in nutritional status within a population under the circumstances in which they now exist influence development. Finally, and related to this last point, because such studies manipulate supplementation and not deprivation, they can only indirectly address the question of how nutritional deficits (i.e., below some agreed-upon normal level) affect development.

Finally, there are the clinical intervention studies. A major methodological concern in these studies, whether or not they involve a control group, is that they generally concern the effects of a psychosocial as well as a nutritional intervention on child development. The question they address is not one of the effects of nutrition per se on behavioral development, but the combined effects of nutrition and stimulation on the behavioral outcome.

It is important to recognize these distinctions in integrating the results of different research studies. Together, the studies provide important information about the role of nutritional deficit and/or supplementation on the child's later development. Discrepant results across studies, however, must be considered with regard to the distinctions between study designs and the research issues they address.

4.9.2 Summary of findings

From the retrospective studies of severely malnourished children we see that, in general, children with a history of undernutrition in the first year or second year of life are more likely to show developmental deficits than children not experiencing such malnutrition when the malnutrition occurs in the context in which it normally occurs—that is, situations of poor material resources, little educational stimulation, and susceptibility to infectious disease. A major contribution of the research on malnutrition secondary to neonatal illness is to sensitize us to the importance of the environmental context of malnutrition. Although this was stressed by the early clinical case studies of malnourished children, the demonstration of normal cognitive development in survivors of serious undernutrition due to intestinal dysfunction when malnutrition was experienced in a protective environment buttresses the position that nutrition and environment cannot and should not be 'disentangled' in considering the threat of malnutrition to the child's development. Malnutrition may pose its most severe threat to development under certain circumstances, and studying malnutrition in its natural context allows us to better understand these circumstances.

The nutritional supplementation studies and the correlational studies on moderate malnutrition have shown that long-standing energy deficit has detrimental effects not only on growth but on the child's cognitive development. There are a number of methodological concerns with both classes of studies and these were elaborated in the previous chapter. A major problem in the supplementation studies concerns the role of environmental stimulation in accounting for behavior outcomes. Other threats to the validity of the studies relate to insufficient attention to the functional form of relationships between nutrient intake and behavior, and incomplete models of the processes mediating such relationships.

In the correlational studies, there are covarying environmental factors which raise questions about the paths by which nutrition has affected behavior. Still, the correlational studies provide an empirical rationale for studying long-term sequaelae of subclinical malnutrition, and indicate the importance of measuring salient dimensions of the extranutritional environment if one is to derive hypotheses about the manner in which chronic undernutrition affects the child's development.

The studies on failure-to-thrive children, intrauterine growth-retarded children, and clinical intervention with severely malnourished children support further the contention that early undernutrition has adverse consequences for the child's development. Each of these categories of studies provides only indirect evidence for nutrition-behavior relationships, since the failure-to-thrive studies often involve a classification which is based in part on evidence of interactional disturbance; the intrauterine growth-retardation studies imply fetal undernutrition but do not necessarily involve restricted food intake on the part of the mother; and the clinical intervention studies generally focus on the effects of stimulation over and above nutritional rehabilitation on recovery from severe malnutrition. These studies have been influential, however, in showing, first, the types of infant behaviors which are affected by poor growth and/or failures in feeding; second, the possible means by which such behavioral disturbances can affect the child's later interactional characteristics and thus affective development; and, third, the importance of stimulation in the treatment of malnutrition. Together the studies suggest that often children who show impaired linear growth or failure to gain weight (due to any of a number of prenatal and/or fetal influences) as newborns have difficulties organizing themselves for attention to the environment and social involvement. Such characteristics pose difficulties for caregivers and affect the nature of the child's social learning experiences in early life. Interventions that help to provide children who have been thus deprived due to malnutrition and/or other physical insult with more stimulating and supportive environments can have long-term ameliorative consequences.

In sum, the studies suggest an interactive model for the long-term effects of early undernutrition, and the potential for recovery through nutritional and environmental manipulations.

5.0 CONCLUSIONS AND IMPLICATIONS FOR THE PRESENT RESEARCH

Chronic malnutrition characterizes much of the world's child population. The effects of severe malnutrition on the structural integrity of the organism are well documented. Impairment of defenses against disease (Neumann, Jelliffe, & Jelliffe, 1978), reduced neural cell growth (Dyson & Jones, 1976), and delayed neural maturation (Dobbing, 1964) are all associated with nutritional deficit in early life. Chronic undernutrition has been associated with reduced stature, head circumference, and physical development (Livingston, Calloway, MacGregor, Fisher, & Hastings, 1975; Underwood et al., 1967).

Less well understood are the behavioral consequences of malnutrition, particularly mild-to-moderate undernutrition. Whereas there is documentation of lasting impairments in cognitive processes in children who experienced severe malnutrition in infancy (McLaren, Yaktin, Kanawati, Sabbagh, & Kadi, 1973; Monckeberg, 1968), the relations between mild-to-moderate malnutrition and later cognitive dysfunction are not strong (see Pollitt & Thomson, 1977, for a review). Further, research has not made entirely clear whether empirical relations between degree of nutritional deficit in early life and later cognitive outcomes remain significant when other important socioeconomic variables (e.g., conditions of housing, stimulation in the home) are controlled (Lloyd-Still, 1976).

Particularly important is the fact that relatively little is known about the effects of chronic malnutrition on children's later social behaviors and emotional characteristics. Few human studies have examined relationships between measures of nutritional deficit in early life and later social and emotional behavior. The few studies that have done so, however, raise provocative issues and suggest that social-emotional well-being may be particularly vulnerable to the effects of early malnutrition.

The studies that have been done in this area have focused on infancy, comparing the behavioral characteristics of infants with a history of undernutrition (including prenatal nutritional deficit) with those of better nourished infants. From the research a consistent pattern of behavioral characteristics of malnourished infants appears, including impaired atttentional processes (Lester, 1975); reduced social responsiveness (Brazelton, Tronick, Lechtig, Lasky, & Klein, 1977; Mora et al., 1979); heightened irritability and inability to tolerate frustration (Mora et al., 1979); and low activity level, reduced independence, and diminished affect (Chavez & Martinez, 1979). Lester (1979) suggested that such effects may occur within a synergistic system where the malnourished infant is less successful at engaging caretakers in interaction and, in turn, is responded to less often and with less sensitivity, resulting in a failure to develop normal patterns of social interaction. If so, we should expect that insufficient nutrition early in life would be associated with poor interpersonal skills and general lack of social responsiveness in later childhood. Thus, the recent research on malnourished human infants provides a theoretical rationale for an investigation of the effects of malnutrition on social-emotional functioning in childhood.

Another perhaps equally compelling theoretical basis for study is the large amount of research on the behavioral characteristics of severely malnourished animals. Experimentally malnourished animals are characterized by apathy, passivity, and inability to sustain attention (Strobel & Zimmerman, 1971). They show reduced exploration and curiosity (Frankova, 1973b), fearful avoidance of new stimuli (Strobel & Zimmerman, 1972), and a failure to respond normally to social initiations from other animals. All of these characteristics would appear to indicate a 'functional avoidance' and a 'functional deprivation' (Barnes, 1976) of new situations and stimuli. Other characteristics of severely malnourished animals are nonpurposive behavior (Frankova & Barnes, 1968), excitability (Barnes, Moore, & Pond, 1970), and unpredictable aggression (Platt & Stewart, 1968).

Thus, although we cannot extrapolate directly from animals to humans, the above findings from the animal research link malnutrition to important functional disorders. Lack of social responsiveness, emotionality, avoidance of new stimuli, and poor behavioral organization comprise a critical complex of behaviors which, if characteristic of malnourished children, are likely to be associated with impaired emotional responding and poor interpersonal skills. As previously indicated, the studies on malnourished infants show reduced responsiveness to persons, lack of initiative, and heightened irritability relative to better-nourished peers. Such characteristics, if they continue to mark the course of early social development, may be expected to impair the quality of the infant's early social interactions and thereby inhibit the development of social competencies and positive emotional characteristics in childhood. In the present study the behavioral implications of chronic malnutrition in school-age children were investigated.

Chapter Four: A Study of Chronic Undernutrition and Child Behavior

1.0 THEORETICAL CONSIDERATIONS

As shown in the preceding chapters, there is a compelling need to study the effects of early malnutrition on later behavior. The research on malnourished animals shows that protein-energy deprivation, both pre- and postnatal, has effects on animal behavior. Attention, persistence, emotionality, and interest in the environment are adversely affected. The human infant literature suggests a similar complex of behavior impairments: reduced activity, loss of interest in the environment, reduced exploration, diminished affect, and lessened social responsiveness.

In considering these findings we asked, What would be the cumulative effects of a history of undernutrition on the developing child? Before suggesting some of the aspects of child behavior which might be influenced over time by chronic nutritional deprivation, let us consider the processes by which malnutrition may have an influence on behavioral development.

It is a truism that malnutrition occurs most frequently in the context of an impoverished environment. There is little disagreement that the malnourished child often has less satisfactory educational opportunities and training, maternal stimulation, and material resources than the more adequately nourished child. These differences were documented in the previous chapter (see Cravioto & DeLicardie, 1972; Richardson, 1976).

It also is something of a truism that malnutrition and environmental deprivation 'interact' to influence physical and behavioral development. Researchers refer to the synergistic relationship between malnutrition and environmental deprivation in influencing child development (Lester, 1979; Martorell, 1980; Rossetti-Ferreira, 1978). The word synergistic refers to a relationship between two or more variables (i.e., conditions or characteristics) such that (a) each influences the other in a reciprocal and ongoing fashion and (b) the combined effects of the variables on some consequent variable are greater than would be predicted solely on the basis of an additive model for their separate effects.

What has not been as carefully explored, however, are the processes by which the covariations between malnutrition (i.e., the organic condition) and environmental deprivation (the external context) occur. Insight into this relationship should suggest some of the behavioral functions that could ultimately be affected by a history of chronic undernutrition.

Several authors have attempted to describe the way in which the condition of malnutrition may lead the child to experience a different set of environmental experiences than he or she would otherwise experience and how, in turn, the affected environment may induce organic and behavioral changes in the child (Lester, 1979; Rossetti-Ferreira, 1978). To understand these models, let us consider the ways in which organic conditions and environmental factors may be empirically associated.

Plomin et al. (1977) define four types of covariation between two potential cause agents. These are active covariation, passive covariation, reactive covariation, and interaction. The ontogenesis of these constructs lies in behavior genetic theory. They were formulated to describe the ways in which a person's genotype and environment may mutually affect one another. They are applicable, however, to the general issue of the relationship between organism and environment. I will consider each of these types of covariation with respect to the question of covariation between malnutrition and environment as they may influence child behavior and development.

Passive covariation would describe a relationship between malnutrition and environment which occurred not because of the causal influence of child (i.e., nutritional) characteristics on environmental characteristics or vice-versa but simply because the conditions that give rise to malnutrition also give rise to certain environmental conditions. As has been pointed out frequently (Pollitt, 1980; Ricciuti, 1981), the child who is born in an environment that leads to undernutrition (e.g., where the mother has low food availability and poor education) is also, by virtue of that environment, likely to be raised under conditions where there is a lack of stimulation and a high prevalence of infectious disease.

Active covariation refers to a relationship which occurs because of the action of the child on the environment. For example, the better-nourished child may explore the environment more than the malnourished child (Graves, 1976, 1978). As a result, he/she encounters a different set of experiences—different persons, objects, places, interactions. In this case, the characteristics of the child (i.e., nutritional status) have actively influenced the environment he/she experiences.

Reactive covariation is the tendency for the organism to elicit responses of a particular kind from the environment, responses which covary with his/her personal characteristics. For example, a very bright child might elicit different treatment from teachers and parents than a less intelligent child. In the present case, the malnourished child may receive less attention from the parents, due to his/her withdrawn characteristics, than the better-nourished child. In fact, Chavez's data from Mexico (Chavez & Yaschine, 1979) showed that fathers responded more positively and enthusiastically to their children as they became better nourished. This response was attributed to the children's more outgoing, attentive, and responsive behavior with improved nutrition.

Interaction refers to the differing effects of a particular organic condition depending on the environment that the organism encounters, or the environment in which the condition was experienced. Examples of interaction between malnutrition and environment as they influenced behavior and development were given in Chapter Three in the review of research on malnutrition in animals (see the section, Research on the Interaction of Malnutrition and Environmental Deprivation). In the human malnutrition literature, perhaps the best example of nutrition-environment interaction is the finding of Richardson (1979) that the effects of early and severe undernutrition in infancy on later intellectual functioning in Jamaican children were stronger in the lower SES rungs than in the higher levels of social environment. Klein (1979) reported similar findings.

The concept of interaction implies that the same social environments may lead to different outcomes for the organism, depending on the characteristics of the child. To date, there are no empirical data on malnourished infants which document such effects. There is, however, compelling documentation of such processes in premature, underweight newborns. Brazelton et al. (1974) have shown that the degree of maternal stimulation which, in the full-term child, will normally elicit happy and enthusiastic behavioral response may elicit a negative avoidance response in the preterm child. Based on the earlier cited research on the behavioral characteristics of full-term, undernourished infants (Brazelton et al., 1977; Lester, 1979), we might expect that a similar environment-organism interaction could occur for the full-term but prenatally undernourished child.

Let us turn now to the models that have been proposed by Lester (1979) and Rossetti-Ferreira (1978). According to these investigators, early malnutrition stresses the central nervous system of the organism and also depletes energy resources. The child is, therefore, poorly equipped to stimulate the caregiver, as shown in the research studies by Als et al. (1976) and Brazelton et al. (1977), and to respond to the caregiver's initiations. Thus, the child is less likely to develop a normal, reciprocal pattern of interaction with the caregiver. If the caregiver also is malnourished, as is often the case, the problem is exacerbated. Under these conditions, the caregiver may not have the attentional and motivational capacity to be sensitive to the child's limitations and to give the child the additional stimulation and support he/she needs. The result is a cycle of dysfunctional responses leading to disturbed caregiver-child interactions, withdrawal on the part of the child, and neglect or rejection (as documented in the Chavez studies) on the part of the caregiver.

Thus far, in this model, we recognize passive, active, and reactive covariation between organism and environment. The passive covariation is represented by the relationship between maternal nutritional status and child nutritional status. The undernourished child is likely to have an undernourished mother, and the mother herself may show behavioral characteristics reflecting her poor nutritional status. There also is active and reactive covariation. The malnourished child elicits different environmental stimulation than the more adequately nourished child, for example, by reduced exploration and, further, may elicit different responses from persons. In addition to reduced attention, the malnourished child may receive negative responses to behavior, for example, due to the frequency or quality of crying (Lester, 1979).

Finally, we recognize in the synergistic model the role of interaction per se. Malnutrition-environment interaction is seen at both the levels of physical and behavioral response. The malnourished child is less resistant to infectious disease; the same environmental agents which would not be dangerous for the healthy child may lead to infection in the malnourished child (Scrimshaw & Behar, 1961). And at the level of the behavioral response, there is the suggestion, from clinical observation and related research, that the malnourished child is less able to assimilate a degree of stimulation which the well-nourished child would be able to respond to appropriately. The capacity to accept and respond to human stimulation may be particularly affected.

We also should recognize the role of the caregiving environment as a mediator of the relationship between malnutrition and behavior. We have noted that the cycle of undernutrition leading to impoverishment of interaction (possibly resulting, in turn, in further malnutrition) is most likely to occur if the social cultural milieu is also impoverished. In a middle-class environment, in which parents are well-educated and families are less vulnerable to social, economic, and health-related stresses, parents may be better able to make efforts to be sensitive to the undernourished child, efforts which could help to prevent the establishment of dysfunctional interaction patterns. Under adverse social-environmental conditions, the caregiver may be less likely to provide the caregiving behaviors necessary for the baby's recovery.

2.0 POSTULATED EFFECTS ON CHILDREN'S SUBSEQUENT BEHAVIOR AND DEVELOPMENT

What might be the effects of these early experiences on the malnourished child? It would appear that disturbed interaction patterns in the first few years would inhibit the development of interpersonal skills and problem-solving capacities in later childhood, and thus adversely influence the child's social and emotional development. In the following section we discuss the nature of the behavioral characteristics which we would expect to be influenced by a history of undernutrition. We first present an approach to conceptualizing social and emotional functioning in relation to chronic undernutrition. We then discuss a methodology for measurement of these characteristics in field research. Then, in the following section, having described the conceptualization and assessment of the relevant behavior variables, we present a report of research on malnutrition and behavior.

3.0 CONCEPTUALIZATION AND ASSESSMENT OF SOCIAL-EMOTIONAL FUNCTIONING*

3.1 Conceptualization

We have stressed that the major findings from the animal and infant studies on behavioral effects of malnutrition are that malnutrition affects the organism's attentional capacities, affect and emotional control, social responsiveness, and activity level. If we consider the developmental implications of such impairments, we should be able to make hypotheses about the undernourished child's future course of behavioral development. More specifically, by considering the ways in which undernutrition might affect the child during infancy, and the way such behavioral effects might influence the child's experiences with his/her caregiving and peer environments, we should be able to arrive at a formulation which identifies dimensions of childhood behavior functioning which might be affected by malnutrition, and then begin to operationalize these variables.

* A version of this material appeared in Barrett, D.E. (1984) Malnutrition and child behavior: Conceptualization, assessment, and an empirical study of social-emotional functioning. In B. Schurch and J. Brozek (eds.), *Malnutrition and Behavior: A Critical Assessment of Key Issues*. Lausanne, Switzerland: Nestlé Foundation for the Study of Nutrition Problems in the World, pp. 280–306. Reprinted by permission of the publisher.

Consider the behavioral dimensions identified: social responsiveness, activity level, affect, and interest in the environment. It seems reasonable to surmise that an infant who shows impairments in any or all of these characteristics becomes a different type of stimulus both to caregivers and peers. We know (from the previously cited work) that undernourished children tend to be responded to less often and with less enthusiasm (by caregivers) than better-nourished children. We would expect that the young child who does not successfully elicit responses from others would learn to withdraw from social interaction situations and would not learn the social skills which are normal in infants who have developed adequate relationships with principal caregivers (see Brazelton, Koslowski, & Main, 1974, for a description of parent-child reciprocal interaction in infancy). Further, we should expect that disruptions of the caregiver-child relationship would influence the child's interactions with peers in the toddler and preschool periods. There is evidence that it is the secure attachment to the mother that provides the emotional base which allows the child to begin to explore and interact with his/her environment (Ainsworth, 1979). The child who has not been successful in the earliest social interactions with the caregiver is likely to fail in interpersonal interactions with peers and nonfamilial adults. The cumulative effect of such interactive failures would be, it is assumed, a lowered self-esteem, avoidance of new persons and situations, and a general inability to handle routinely stressful social situations, including group problem-solving situations. In addition, we should expect affective withdrawal and lowered physical activity.

Thus, at the childhood stage, we can begin to identify specific dimensions of behavior which could be influenced by the child's nutritional history. It is useful to discuss child social-emotional competence in terms of two general types of characteristics. The first is the child's general social interaction characteristics (i.e., across situations). These include: (a) interaction with peers; (b) behavior toward adults; (c) response to the physical environment; (d) affect, and (e) activity level. Second are the child's characteristic responses to specific types of situations and, in particular, to routinely stressful or pressured situations. These would include responses to: competition; frustrating problems; seeing another person in distress; an unfamiliar environment; and demands to control impulses (e.g., situations calling for delay of gratification or resistance to temptation).

This formulation has heuristic value for two reasons. First, it encompasses almost all of the social behavioral characteristics which have been considered under the rubric social competence (see Zigler & Trickett, 1978). Second, and perhaps more importantly, each of the categories can be conceptualized in terms of specific aspects of function which we would hypothesize to be related to undernutrition. For example, consider the five broad categories of general social interaction characteristics. For each of these categories, we can think of a number of variables which should be sensitive to differences in nutritional history. For example, for interaction with peers, one would predict less frequent social initiations toward others, less sensitive responding to cues from others, more passive or submissive behavior, and less physically active play in undernourished children. In terms of responses to adults, we would expect children with a history of undernutrition to show more dependent behavior and, in particular, more attention seeking. With regard to the response to the physical environment, we might expect less adequately nourished children to show less sustained involvement with materials and with group activities, more wandering, greater susceptibility to distraction from task-irrelevant events, and less careful handling of materials. For affect, we would predict less outward show of expression (happy, angry, concerned), that is, generally flat affect, and more frequent emotional outbursts. Finally, we would expect less well-nourished children to have difficulty modulating activity, with the result being a generally low activity level with periods of uncontrolled activity.

In structured situations which involve some type of routine stress or challenge, the following could be predicted: In group situations involving competition, we would expect lower levels of participation and perhaps a tendency to withdraw when the situation becomes most competitive. In group problem-solving situations (e.g., a puzzle or real-life dilemma), we might predict a tendency to give up easily and to fail to attempt a variety of strategies. For response to another person in distress, we would expect a tendency to fail to notice the other's cues (e.g., facial expression), more uncertainty when (or if) distress cues are noticed, and less overt helping. For response to an unfamiliar environment, we would expect less curiosity (fewer materials explored and lower levels of involvement), and a tendency to withdraw from the situation. Finally, in impulse control situations, we would expect a general inability to inhibit rapid responses—for example, a difficulty in delaying gratification or resisting an opportunity to obtain an immediate reward in the face of a rule or prohibition.

In summary, a useful approach to the conceptualization of social-emotional functioning is to identify general categories of behavior which encompass the most salient features of children's activity and which include dimensions of behavior which are relevant to our hypotheses about the behavioral effects of undernutrition. It is suggested that we consider both situation-general and situation-specific aspects of behavior, and that within each of these general categories we identify specific classes of behavior function for assessment.

3.2 Assessment

Having identified the dimensions of child behavior in which we are interested, the next step is to operationalize the behavior variables. The approaches which may be used to collect and reduce data were discussed in detail in Chapter Two, and their respective advantages and limitations noted.

In our research, we identified classes of behavior relevant to malnutrition which fell into the two categories, situation-general and situation-specific behaviors. The approach we used for the assessment of the situation-general behaviors was first, using a time-sampling procedure, to measure each behavior of interest in a variety of situations and obtain an average frequency count of the behavior per specified time interval (i.e., behavior rate) across situations. We later asked raters to make global assessments on a number of behavior dimensions based on their overall impressions of the children (i.e., after having observed the children in a variety of activities). The dimensions considered in the ratings were to some extent redundant with the behavior categories used in the time-sampling observations. To obtain scores on the situation-specific variables, we created structured situations for the subjects and observed their responses, making behavior ratings on the spot. Thus, although, in general, the assessments took the form of ratings rather than frequency counts, the approach was similar to the time-sampling approach in that we recorded behaviors at the moment of observation and in the situation of interest.

Below we introduce the assessment procedures. In a later section (4.3.4), we describe in greater detail the measurement situations and coding methods. In the section in which we present the results of the research (4.4), we will refer again to the variables described here.

3.2.1 General observation paradigm

Children are observed in social interaction in six-person groups. The assessments cover a period of two days. On each day, the children are observed in a number of situations which provide the opportunity to code both their interpersonal behaviors and their nonsocial behaviors on a time-sampling basis. The activities include clay construction activities, painting a mural, free play in a relatively new environment (after an initial familiarization period), and play with 'physical' materials such as balloons, balls, large blocks, and wagons. In addition, the children are observed in a number of problem-solving situations in which we examine their responses to particular problems. (These are described below, under Situation-specific Assessments.) All situations are organized so that together they comprise a coherent sequence of gamelike activities.

3.2.2 Time-sampling assessments

Across the variety of activities, each child is observed for a series of two-minute periods that are divided into 30-second intervals. Behaviors are coded in five classes: Interaction with Peers, Contact with Adults, Behavior with Physical Environment, Level of Activity, and Emotions. The specific categories are shown in Table 4–1.

For each behavior category, each child receives a score representing the average number of occurrences of that behavior per two-minute interval. Thus, we obtain, for each child, a set of

TABLE 4.1

Behaviors coded on time-sampling basis in small-group activities

Behavior	Reliability
Interaction with peers	
Friendly interaction	.89*
Helps, shares, comforts	.24*
Physical aggression	.50*
Verbal aggression	1.00*
Rough and tumble play	.81*
Intrudes, interferes	.79*
Dominates, asserts self	.81*
Seeks help	.68*
Submissive	.65*
Defends self	.72*
Positive response to bid	.88*
Negative response to bid	.95*
No interaction	.90*
Contact with adults	
Seeks attention	.59*
Seeks physical contact	.00
No interaction	.63*
Behavior with environment	
Involved in group activity	.88*
Plays with materials	.89*
Destructive with materials	.00
Wanders	.95*
Physical play alone	.00
Imaginative play	.00
Distracted by outside events	.84*
Only watching	.92*
Inactive	.79*
Level of activity	
Low	.67*
Moderate	.70*
High	.72*
Very high	.74*
Emotion	
Happy, laughs, smiles	.84*
Anxious	.85*
Angry, hostile	.53*
Sad	.00
Loses control	.00
No emotion shown	.79*

Note. Reliabilities are estimated by intraclass correlation coefficients. See Footnote 5 for a note on the interpretation of zero-level intraclass correlations.
* $p < .001$.

From Barrett, D.E., Radke-Yarrow, M.R., and Klein, R.E. (1982) Chronic malnutrition and child behavior: Effects of early caloric supplementation on social and emotional functioning at school age. *Dev. Psychol.* 18: 541–556.

scores representing the rate of each of a number of theoretically relevant behaviors across a variety of situations.

3.2.3 Global ratings

Following the two days of observations, each observer rates each child on a number of behavior dimensions which we have hypothesized (based on the theoretical considerations described in the first section of this chapter) to be related to undernutrition. On scales of 1 to 5, subjects are rated for energy level, work orientation, attention (to the group activity), communication skills, disposition, obedience to adult directions, and self-confidence.

3.2.4 Situation-specific assessments

Data are collected on the child's responses to the following situations: (a) novel environment (unfamiliar toys, persons, setting); (b) competitive game (a relay race involving popping balloons); (c) frustrating problem (unopenable container containing a prize), and (d) an impulse control game (like Simon Says but without visual cues). More detailed descriptions of these situations can be found in this chapter under Report of the Study. (See section 4.3.4, Dependent Variables.)

The variables coded in these situations reflect our hypotheses about the behaviors differentiating children with better and poorer nutritional histories. They are:

1. Novel Environment
   a. Latency to first contact with toys (rating of 1 to 4)
   b. Number of materials touched (number from 0 to 5 or more)
   c. Level of involvement with individual toys (rating of 1 to 3)

2. Competitive Game
   a. Level of task involvement (rating of 1 to 3)
   b. Positive affect (rating of 1 to 4)
   c. Sense of competition (presence or absence)
   d. Fearfulness about popping balloons (rating of 1 to 3)

3. Frustrating Problem

   a. Number of strategies used to solve the problem (number from 0 to 7)
4. Impulse Control Game
   a. Number of errors of commission
   b. Number of anticipatory responses
   c. Number of errors of omission

3.2.5 Evaluating observer agreement

To obtain information on observer agreement, we instruct each of three observers to watch three children per activity. In the time-sampling assessments, children are watched one at a time, in sequence, for two-minute periods. Since there are three observers and six children, we are able to obtain matching observational records and thus determine interrater agreement (on a given behavior). The data are analyzed for observer agreement using the intraclass correlation (Bartko, 1966) which indicates what percent of variance in scores for a group of subjects is due to 'subject' variance as opposed to 'rater' variance. For global ratings and for ratings of responses to specific situations we compute coefficient weighted kappa (Cohen, 1968), which indicates the degree of agreement between raters, correcting for chance agreement. Weighting is linear; for example, a disagreement of two scale points is weighted twice as heavily as a disagreement of one point. (Intraclass correlations and kappa statistics obtained in our research are reported in the following section of this chapter.)

We have presented an approach to the conceptualization and assessment of social and emotional functioning in examining relationships between malnutrition and child behavior. We may now ask, Is there empirical evidence that this can be a fruitful approach for studying the effects of malnutrition on children's behavior? To address this question, we summarize the results of the first author's recent study on the effects of undernutrition on children's behavior at school age (Barrett, Yarrow, & Klein, 1982; Barrett & Yarrow, 1985).

4.0 REPORT OF THE STUDY

4.1 The Study Population

The childen in the present study were from three villages in Guatemala. The children in these villages had been participants in a longitudinal study on nutrition and behavior. The study was carried out between 1969 and 1977 by the Institute of Nutrition of Central America and Panama (INCAP). The purpose of the study was to determine whether there was a causal relationship between chronic undernutrition and physical and cognitive development. The method used was to provide nutritional supplements to an endemically malnourished population and to examine the relationships between the amount of supplement intake and later development. If a high degree of supplementation of nutrients limiting in the daily diet was related to optimal behavioral and developmental outcomes, a causal influence of nutrition on behavior could be inferred.

How poorly nourished were the children prior to supplementation and what nutrients were limiting in their diets? Estimates of mean energy and protein intake showed that at all ages, mean daily energy intake for the study population was substantially below estimated daily requirements. These deficits were particularly large for preschool children and for mothers. For example, 4-year-old children at the beginning of the longitudinal study received approximately 80 kcal/kg/day, compared with estimated needs of 99 kcal/kg/day (FAO/WHO, 1973). Mothers during pregnancy received about 1300 calories per day, only about one-half the recommended allowance.

In addition, there were anthropometric data showing the degree of early nutritional deficit. The average height of preschool children in the sample was at less than the fifth percentile of the Denver norms. The average height of each parent also was at less than the fifth percentile. Average weights for height generally were higher, however, in the 25th to 50th percentiles for each age group.

Finally, there were the data on weight for age. Using the Gomez et al. (1956) criterion of 75 to 90 percent of expected weight for age as marginally malnourished, baseline data obtained at the beginning of the INCAP study showed that over 80 percent of the school age population was at least mildly to moderately undernourished. In adults as well, weights indicated suboptimal nutrition.

To better complete the picture of the study population, demographic and descriptive data are necessary.

Engle (Engle et al., 1983) has provided a description of the villages. Participants in the study were from families engaged primarily in subsistence agriculture. The main crops raised are corn and beans. There is little migration to or from communities.

The median income is approximately $500 per year (as of 1975). Nearly all houses lack sanitary facilities. Most houses are two-room houses made of adobe or other local materials. One room serves as the kitchen and the other as sleeping quarters for the family. Clothing often is lacking. Many families have no sweaters or shoes. Pictures of the villages and the children appear on pages (210–215).

A major characteristic of the study villages is the extremely high rates of morbidity and mortality. Baseline data on morbidity were not collected, but data were collected throughout the course of the study. While they underestimate the degree of illness which would have been recorded were there no nutritional intervention, they are, nevertheless, very powerful evidence of the disadvantage of the population.

The incidence of respiratory infections, diarrhea, and anorexia was examined. Percent days ill from birth to 3 years was documented. Average incidences for the three types of illness were extremely high in infancy. For example, average percent days ill with diarrhea for boys from birth to 2 years was 12.5, for anorexia it was 21.6, and for all illness, 28.1. For girls, the figures were 11.9, 20, and 26.3, respectively. Thus, almost one-fourth of the time children were observed, they were ill with respiratory illness, anorexia, or diarrhea (INCAP, pers. comm.). The mortality rate for newborns was 12.1% for low-birthweight infants (<2.5 kg), and 6.7% for infants 2.5 to 3.0 kg (Habicht et al., 1976). Engle provides the following additional information. A total of 46 percent of mothers in the sample reported having at least one child die. A total of 85 percent of child deaths occur within the first week of life; 2.4 percent after age 3 years.

Top: Village of Santo Domingo, County of El Progresso, Guatemala. *Bottom:* House in Santo Domingo.

Photos: David E. Barrett

Top: Older girls playing in the central square in Santo Domingo. *Bottom:* School in Santo Domingo.

Top: Observing children in San Juan, El Progresso, Guatemala. *Bottom:* Village of Conocaste, El Progresso, Guatemala.

Top: Group of boys in Santo Domingo. *Bottom:* Group of girls in Santo Domingo.

Top: Group of boys in Santo Domingo. The young woman in the photo helped us by bringing the subjects to and from their houses. *Bottom:* Group of girls in Santo Domingo.

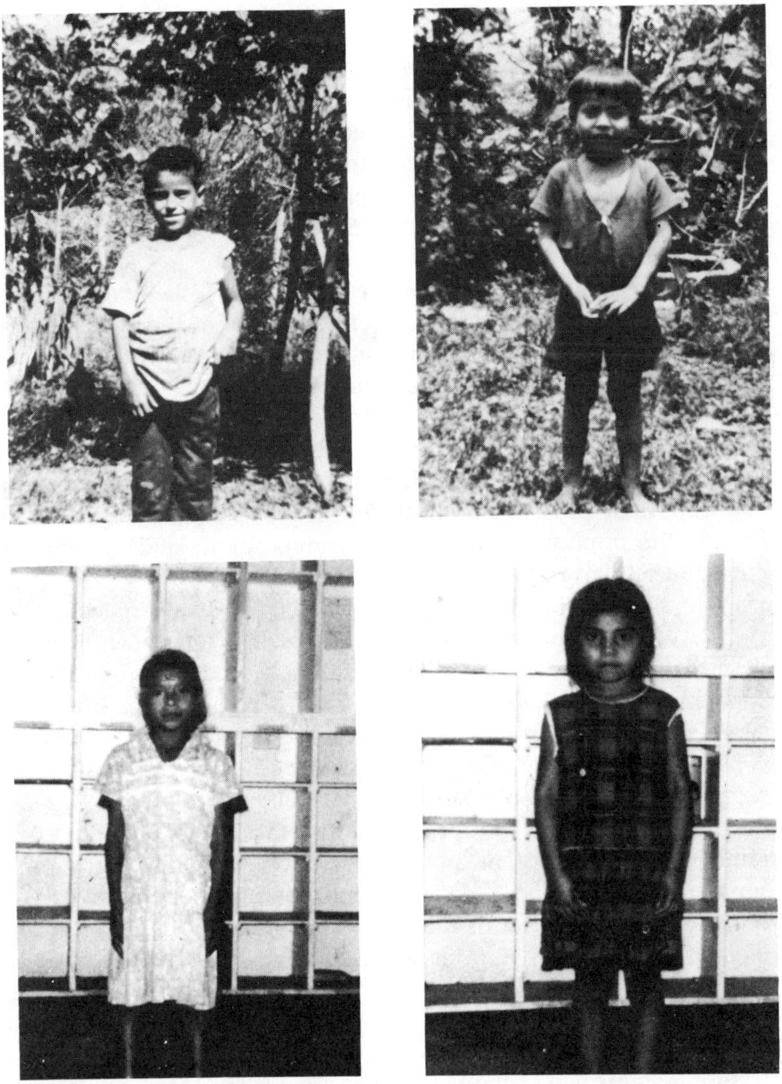

Top: Two boys from Conocaste. The boy on the left is 6 years, 9 months old, and the boy on the right is 7 years, 9 months old. Both are average weight for height but stunted. *Bottom:* Two girls from San Juan. The girl on the left, age 8 years, is of average weight for height but is stunted. The girl on the right, age 7 years, 9 months, is both wasted and stunted.

4.2 The design of the study

The present study uses a quasiexperimental design. Villages were assigned at random to two different treatments. One treatment group received a protein-calorie supplement. The other treatment group received a calorie-only supplement. Subjects in the two groups were compared on measures of later growth and development.

In the original INCAP study (see Klein, 1979), the investigators wished to determine the relative effectiveness of the two supplements as a means of investigating the role of protein-calorie malnutrition on later development. The rationale was that if differences in protein in the two groups were related to differences in growth and devlopment, that would show the effects of protein on human growth and development. If individual differences in calorie and/or protein intake *within* groups were related to outcomes, the importance of these nutrients for later development could be similarly assessed.

Four study villages were selected. The four were from the county of El Progresso, Guatemala, and had the characteristics described previously. Two villages were assigned a protein-calorie supplement, called *atole*; two villages were assigned a calorie-only supplement called *fresco*. The protein-calorie supplement (atole) consisted of a vegetable-protein mixture, dry skim milk, and sugar. The calorie-only supplement contained only sugar and flavoring. One cup (180 ml) of protein-calorie supplement contained 11.5 g of protein and 163 kcal. One cup of calorie-only supplement contained 59 kcal. Both preparations included essential vitamins and minerals that were thought to be limiting in the diets of the study populations. Each supplement contained the following vitamins and minerals: ascorbic acid (4.0 mg/cup); thiamine (1.1 mg/cup); vitamin A (1.2 mg/cup); iron (5.0–5.4 mg/cup); and fluoride (0.2 mg/cup).

Supplements were administered at central food stations in each study village. Attendance was voluntary. Daily supplemental caloric intake was recorded (kcal/kg/day) for all women during pregnancy and for all children under 7 years of age.

We can now ask: What are the features of this design that allow us to address the important questions of malnutrition and behavior?

What questions are not directly addressed but receive indirect attention?

The present design considers intake of calories as the independent variable. In the original design, the interest was in the effects of proteins versus calories in promoting physical growth and psychological development. Over the course of the study, however, a different perspective emerged. INCAP investigators (Engle et al., 1979; Klein, 1979) consistently found that supplement calorie intake was the best predictor of physical growth and cognitive development up to the preschool years, when compared to other nutritional variables, specifically, diet or supplemental protein.

Some of the specific findings were the following. First, maternal calorie supplementation during pregnancy was associated with reduced infant mortality and reduced incidence of low birthweight. When mothers were compared for amount of caloric supplementation during pregnancy, the risk of low birthweight babies among low-supplemented mothers was almost twice as high as the risk for high-supplemented mothers (Lechtig et al., 1975b). Second, there were significant associations between amount of caloric supplementation for the mother and stillbirths, and with infant deaths during the first, second, third, and fourth quarters of the first year of life. Risk of infant death at birth to 3 months, for example, was increased from 2.5 percent to about 5 percent; that is, by almost 100 percent, for low-supplemented women.

Third, there were significant associations between caloric supplementation during infancy and child growth during infancy and the preschool years. For example, our data show that caloric supplementation from birth to age 2 predicted height at age 4 ($r = 0.39$; $p < .005$ for boys, and $r = 0.27$, $p < 0.05$ for girls). In addition, and perhaps more importantly, high caloric supplementation reduced the differences between mean heights of preschool age children in Guatemala compared to United States reference populations. When children are divided according to level of calorie supplementation into categories 0 to 50 kcal day, 50–200 kcal/day, and greater than 300 kcal/day (estimated energy needs for the child at 15–36 months are approximately 300 kcal/day), differences between the groups in attained growth is striking. For example, the difference in height between children consuming more than 200 kcal/day and those consuming less than 50 kcal was

3.5 cm for boys and 3.8 cm for girls. This difference in size is almost one-third the difference between the average height of the chronically malnourished Guatemalan child and the United States standard.

Fourth, there were significant associations between supplemental caloric intake and children's cognitive development. There were strong relations between amount of supplementation and psychological test performance beginning at 15 months of age (Lechtig et al., 1975a) and up until preschool (Klein et al., 1977).

The consistency of the calorie supplement findings across villages and type of supplementation suggested the use of caloric supplementation as the primary measure of nutritional history.

4.2.1 Theoretical rationale for the design

In retrospect, the use of supplement calories as the independent variable made theoretical sense. As indicated previously, the children in the study were chronically energy deficient, with daily intakes well under estimated requirements. Baseline data on diet protein intake, however, indicated that for children at all age levels, daily protein intake was at or above both required and safe levels.

From a physiological standpoint, this energy deficit means that children were likely to burn protein to satisfy energy needs, thus threatening the available store of body protein. The hypothesized effect of supplemental calories would be to spare body protein by providing adequate nutrients for energy needs. Thus, low supplemental caloric intake in a supplemented population would represent relative energy deficit and risk for retarded growth and development.

We have cited empirical evidence supporting this interpretation, specifically, the relationships between caloric supplementation and mortality, birthweight, physical growth, and cognitive development. Further, Martorell et al. (1976) reviewed the many studies on the effects of protein and/or calorie supplementation on postnatal physical growth in developing countries. They concluded that in the experiments reviewed, protein-calorie supplementation was causally related to physical growth. The relative importance of proteins versus calories depended on which nutrient was limiting in the home diet. When calories but not protein were limiting, calories alone seemed to improve growth rates. Thus, on both theoretical and empirical grounds, level of caloric supplement intake appears to be an appropriate index of nutritional status for the study population.

4.2.2 *Methodological considerations: Diet vs. supplementation*

One of the questions posed by the present design is, Since diet calories were known to be deficient in the population, and since caloric supplementation was known to have an effect on growth, why not use individual differences in diet calories as the major independent variable? That is, why not look at the relationship between diet calories and later physical, cognitive, social, and emotional outcomes instead of looking at the effects of a relatively small (i.e., compared to daily diet) dose of supplement?

There are three reasons why caloric supplementation was the more appropriate measure of relative nutritional adequacy. First, there was the measurement consideration. Reliability of supplement intake was substantially higher than reliability of diet intakes. Second, there was the major methodological issue. Daily diet was confounded with home environment variables, while supplement administration approximated a random treatment. Third, there was the logic of boosting children from a deficient to an adequate calorie level. Theoretically, mothers could provide 200 to 400 additional calories per day to themselves and/or their children (depending on which supplement they took). If the assumption is that original diets were deficient in calories and that supplementation would boost calories to more adequate levels, individual differences in supplementation would be a reasonable choice for an independent variable. Let us consider these three issues.

4.2.2.1 Measurement properties of diet and supplementation It is well known that dietary recall surveys are subject to a relatively large measurement error when applied to individuals (i.e., vs. groups). For example, Habicht reports (Habicht, Yarbrough, & Klein, 1974) that when 24-hour home recall surveys were used to record home protein intake, three-month test-retest correlations were on the order of 0.51. This means that the maximum correlation of home protein intake with any other variable could only be 0.71. That is, because of the intrachild measurement error for home protein intake, the variable could only account for approximately 50 percent of the variance in any criterion variable—for example, growth or cognitive development. In contrast, supplement protein and calorie intake was measured with near-perfect accuracy. Obtained daily (and, thus, without a need to estimate intake over a larger period) supplemental nutrient intake was measured by hand and without any discernible error (e.g., as could be determined by interobserver disagreement or discrepancy). Thus, the maximum correlation with a criterion variable would be limited only by the unreliability of the criterion variable.

Habicht et al. (1974) presented the hypothetical correlations between home protein intake measured by 24-hour diet recall and supplemental protein intake with change in height over a period of six months (between 2-1/2 and 3 years of age) as the criterion. Reliability of changes in height was approximately 0.64. The maximum correlation between supplement protein and changes in height was 0.81, while the maximum correlation between diet protein and changes in height was 0.58.

4.2.2.2 Correlations with environmental variables A second consideration is the relationship between diet, supplementation, and other environmental variables. One would expect that differences in nutritional practices in the home would reflect a number of different social-economic factors. In contrast, a purpose of the supplementation was to create differences that would be unrelated to such characteristics. The measure which we have used in the research to represent the general home living conditions was called SES (see Klein, 1979), and represented a summated rating of (a) the quality of the house—floor, walls, ceiling, and location of kitchen; (b) amount of parent stimulation of preschool children in the home; (c) shoes and sweaters. Our analyses showed that measures of child supplementation and also mother supplementation were unrelated to this general index of social status, for both sexes and at all ages at which supplementation was assessed. In contrast, in several instances, measures of diet intake were related to socioeconomic status. For example, diet calories for the child estimated for age 2 was positively associated with SES for boys, and this despite the somewhat high measurement error on the nutritional measure.

Although nutritional supplementation was not significantly related to social environment as we measured it, there are other concerns about using nutritional supplementation as a measure of malnutrition—specifically, (a) motivational factors related to use of the intervention, and (b) confounding of nutritional with nonnutritional stimulation. These are addressed below under the heading, Construct Validity of the Intervention.

4.2.2.3 Adequate levels of nutrient intake The third consideration involves the issue of threshold level of calorie intake for appropriate development. One of the assumptions in choosing this study population was that home diet was inadequate to meet growth needs. While individual differences between children receiving more and less calorie-rich home diets might predict individual differences in growth and development, it seemed reasonable to believe that the greatest differences would be found between persons receiving inadequate and adequate caloric levels. That is, it seemed reasonable to assume that greater differentiation would be achieved by contrasting those whose supplementation was inadequate to bring them to an optimal or adequate level of energy intake, with those whose intake was high enough to do so. INCAP investigators have consistently found that contrasts between mothers with nonoptimal versus adequate supplemental caloric intakes have proved most effective in predicting later developmental outcomes. For example, Lechtig et al. (1975a,b) found that contrasts between mothers receiving less than 20,000 supplemental calories over the course of pregnancy versus greater than 20,000 supplemental calories predicted frequency of optimal birthweight infants and percentage of children who were not developmentally delayed at 15 months on measures of sensorimotor and cognitive functioning.

4.2.3 Construct validity of the intervention

We noted in the previous chapter that a major question in nutritional intervention research is, Was the nutrition per se responsible for the experimental outcome? Or, more broadly, Was the effect that was identified due to the causes that we believe brought about that effect? In the present case, two alternative hypotheses arise: First, that the effect of nutritional supplementation was not due to something in the intervention but was related to pre-existing characteristics of the mother and/or child. This is a possibility because mothers were free to consume as much supplement of a given type as they wished. A second alternative is that the effect of the nutritional intervention was due to the intervention but was not specifically due to the nutrition component.

The possible confounding of maternal personality factors with supplementation has been addressed by INCAP investigators (Engle et al., 1979). It obviously is the case that mothers with certain backgrounds, needs, or desires for their children might make better use of the supplementation program than others. For example, better-educated mothers might be expected to take more supplementation than less well-educated mothers. Neither the original INCAP analyses of the total sample data nor the present analysis of a subsample have identified any maternal personality or demographic variables related both to the mother's or child's use of supplementation and the child behavioral outcomes. Relevant analyses are given later, in the Results section. Thus, there is no evidence that supplementation effects are due to non-nutritional, environmental, or maternal factors.

There is, however, a suggestion in our data that maternal expectations may have influenced their use of supplementation. We found, for boys, a marginally significant correlation between the child's birthweight and the degree to which the mothers provided supplementation during the first two years of life ($r = .23$, $p < .08$). Gestational age was also marginally associated with child supplementation in the first two years ($r = .22$, $p < .06$). For girls, there were similar findings with respect to child supplementation for ages 2 to 4 years. These findings suggest the possibility that mothers may have differentially supplemented those children who are healthier.

The second problem of construct validity is the one I referred to in Chapter Two, the fact that a nutritional intervention typically involves general stimulation or enrichment for the participants. For example, in the present study, each time the mother came to the supplementation center she would see other mothers, receive personal attention from the INCAP staff, and have the opportunity to engage in informal conversation. Certainly, such an experience could contribute to a person's feeling of self-worth and/or sense of belonging to a community. Without measurement of such effects, we can do no more than speculate. It seems reasonable to assume, however, that since amount of supplementation was correlated with frequency of attendance at the centers, nutritional effects on child development may have been buttressed by positive effects on mother attitudes and behavior, which may then have had a direct effect on interaction with the child.

4.2.4 Data analysis: General approach

The analysis approach was to look at the implications of individual differences in caloric intake for later behavioral functioning. Two different strategies were used. One involved looking at the relative importance of supplementation at different times for behavioral development. We determined the amount of supplementation to the mother and/or child at different times of development, created continuous variables representing each of these supplement inputs, and examined the cumulative impact of supplementation at different developmental periods for the child's later development. Note that the assumption of this approach is that by seeing at what periods supplementation had its strongest impact on behavior, one might be able to make inferences about the importance of nutrition at particular periods for later behavioral development. The second approach treated supplementation as a categorical variable. Based on explicit criteria to be described later, children were classified as more or less adequately supplemented. We examined the implications of high and low supplementation for later developmental outcomes.

4.3 Method

4.3.1 Subjects

Subjects were 78 boys and 60 girls from the villages of Santo

Domingo, Conocaste, and San Juan in the county of El Progresso, Guatemala. Children were between the ages of 6 years 1 month and 8 years 3 months at the time of the data collection. Villages were represented in the study in approximate proportion to village size: There were 60 subjects from Santo Domingo, 54 from Conocaste, and 24 from San Juan. These three agricultural communities are located 30–60 kilometers northeast of Guatemala City. The adult literacy rate is approximately 50 percent. The study population has been identified as chronically malnourished: 80 to 85 percent of preschool-age children were classified as at least mildly to moderately undernourished in INCAP's baseline survey according to Gomez et al.'s (1956) criterion of weight for age.

In selecting the sample, an attempt was made to ensure adequate variability in nutritional history. Therefore, supplemental caloric intake of mother and child was used as a selection variable. On the basis of INCAP's longitudinal data, all children in the villages were assigned to 'good,' 'average,' or 'poor' classifications with respect to caloric supplementation of the mother during pregnancy and of the child from birth to 4 years.*

Because height for age has often been used as an index of degree of nutritional deprivation (Brozek, 1979), children also were classified as 'tall' (third or fourth quartile) or 'short' (first or second quartile). The sample was then selected so that (a) there would be approximately equal numbers of well-supplemented and poorly supplemented children and equal numbers of tall and short children, and (b) within villages, the characteristics of the sample with respect to supplementation history and height would reflect the

*Children were assigned scores of 1 to 3 for mother caloric supplementation history and scores of 1 to 5 for child caloric supplementation history. For mother supplementation history, scores were assigned as follows: 1 = less than 5,000 supplemental calories during pregnancy; 2 = 5,000–20,000 supplemental calories during pregnancy; 3 = more than 20,000 supplemental calories during pregnancy. For child supplementation, scores were based on the percentage of three-month periods in which the child was poorly supplemented (less than 5,000 calories) or well supplemented (more than 10,000 calories). Ratings were as follows: 1 = poorly supplemented in more than three-fourths of the periods; 2 = poorly supplemented in more than one-half of the periods; 3 = poorly supplemented in less than one-half of the periods, but not well supplemented in more than one-half of the periods; 4 = well supplemented in more than one-half of the periods; 5 = well supplemented in more than three-fourths of the periods. Combinations of mother and child supplement ratings were used to classify children as good, poor, or average in supplementation.

village population. The final sample of 138 children included 56 with good supplementation, 29 with average supplementation, and 53 with poor supplementation. A total of 68 children were tall, and 70 were short.

4.3.2 Independent variables

In our study, there were two classes of independent or predictor variables: those used in the cumulative prediction analysis and those used in the group comparison analysis.

The variables used in the cumulative prediction analyses represented (a) amounts of supplemental calories ingested by the mother during pregnancy and by the child at different periods of infancy or early childhood, and (b) a measure of socioeconomic status representing quality of home living conditions, described previously.

The supplemental calorie measures were (a) total supplemental calorie intake of the mother during pregnancy, (b) total supplemental calorie intake of the child from birth to 2 years, and (c) total supplemental calorie intake of the child from 2 to 4 years. These variables were constructed from the original INCAP data. INCAP had recorded total supplemental calorie intake of the mother and child over consecutive three-month periods (or longer periods after age 3 years) over the course of the study.

It is important to recognize that a number of other measures of child health and development were obtained by INCAP and were available as independent variables. For example, data were available on child illness (anorexia, respiratory illness, diarrhea), anthropometry (height, weight, head and arm circumference, and various skinfold measurements), and radiology at various points in time over the course of infancy and early childhood. Data also were obtained on maternal and paternal anthropometry, education, and literacy. In addition, there were the data on dietary intake of calories and protein, and supplemental protein intake.

The rationale for focusing on supplemental calories has been given: The population was calorie-deficient, calorie supplementation was measured precisely, and earlier analyses indicated significant effects of calorie supplementation on linear growth and early cognitive development.

4.3.3 Data analysis procedures

4.3.3.1 Multiple regression analyses The three supplementation variables were used as predictor variables in multiple regression analyses, controlling for socioeconomic status, to predict the behavioral variables. For each dependent variable of interest, multiple regression analyses were carried out to examine the individual and cumulative predictive ability of the supplementation variables. The first analysis was conducted across villages. Socioeconomic status was entered into the equation, followed by supplemental calories of the mother during pregnancy, supplemental calories of the child from birth to 2 years, and supplemental calories of the child from ages 2 to 4 years. The analyses thus addressed these questions: (a) How well can we predict individual differences on the dependent variable on the basis of prenatal supplementation, having controlled for home environment? (b) How much of the variance in the dependent measure can we account for on the basis of supplementation during infancy, having accounted for the effects of SES and maternal supplementation? (c) How much further can we account for variance in the dependent variable, if we now include a measure of supplementation in the child's third and fourth years? It should be noted that the same analyses also allowed us to determine which of the supplementation measures was the best predictor of the outcome measures, controlling for SES (see Nie, Hull, Jenkins, Steinbrenner, & Bent, 1975, pp. 358–359).

In a second set of analyses we conducted multiple regression analyses again on each dependent variable within each of the two larger villages, one a protein-calorie village (Conocaste) and one a calorie-only village (Santo Domingo). In these analyses, independent variables were entered in the same order as they were entered in the first, across-village analyses.

4.3.3.2 Analysis of variance In the group comparison analysis (i.e., poorly supplemented vs. adequately supplemented), two dichotomous variables were used in a 2 X 2 analysis of variance: sex and level of supplementation.

We classified children as high or low in supplementation history based on their relative standings on each of the three supplementation variables described in the preceding section. Children were classified Low Supplementation if their total supplemental calorie intake during any of the three periods (prenatal, birth to 2 years, 2 to 4 years) was below one standard deviation below the mean for that period. All other children were classified High Supplementation. There were 69 subjects in each category. We compared High versus Low Supplement children on the criterion measures.

Children in the high and low supplementation groups received averages of 12,626 calories over pregnancy for high supplementation, 2,453 calories for low supplementation; 81,214 calories over the period birth to 2 years for high supplementation, 7,268 calories for low supplementation; and 12,626 calories over the period 2 to 4 years for high supplementation, and 3,308 calories for low supplementation. The groups did not differ in literacy, SES, maternal age (mean = 33 for low supplementation, 32 for high supplementation), or number of years that the mother attended school (mean = two years). The low supplementation mothers had passed more grades in school than high supplementation mothers (two years vs. one year), contrary to expectation. This variable was not predictive of child behavior (see Results section).

4.3.4 Dependent variables (behavior measures)

Children were observed in social interaction in six-person groups. In addition, each child was tested on a psychological battery. The children were brought to a central location in the village for the observations and tests. Assessments took place within a two-day period. There were approximately five hours of small-group observations per child, two and a half hours on each of the two mornings. Afternoons were used for individual testing. There were three observer-testers.

4.3.4.1 Group observations In each six-person group, children were of the same sex, were within 1 year in age when possible, and were of varied statures and supplementation histories. Male and female groups were observed alternately in each village. In each of the two small-group sessions, children were observed in approximately nine different situations. The situations were joined together so that they comprised a smoothly running sequence of gamelike activities. The paradigm was adapted from Yarrow, Scott, and Waxler (1973).

In the situations in which children interacted freely without a specific group goal objective, an ongoing record of each child's activity was obtained using a time-sampling procedure. As described in the preceding section, each child was observed for a series of two-minute periods divided into 30-second intervals. The coding categories are shown in Table 4–1, along with the observing reliabilities. Each child was observed for approximately 25 two-minute periods. Each child's score is the mean frequency of that behavior per two-minute interval.

Interspersed between the unstructured free play situations (e.g., painting a mural, playing with toys, clay construction) were the situations in which we examined the child's responses to particular problems and/or frustrations. The situations are illustrated in Figures 4–1 through 4–4 and are described below.

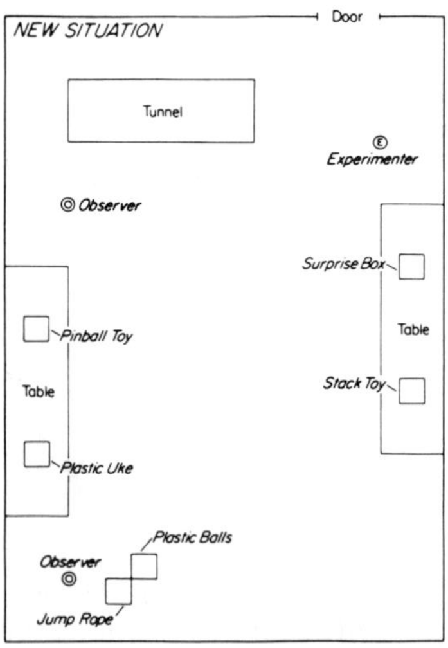

Figure 4-1. Schematic illustration of new situation.

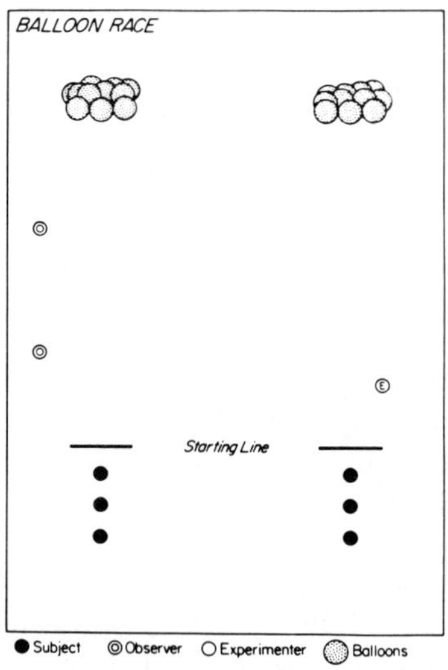

Figure 4-2. Schematic illustration of balloon race (competitive game).

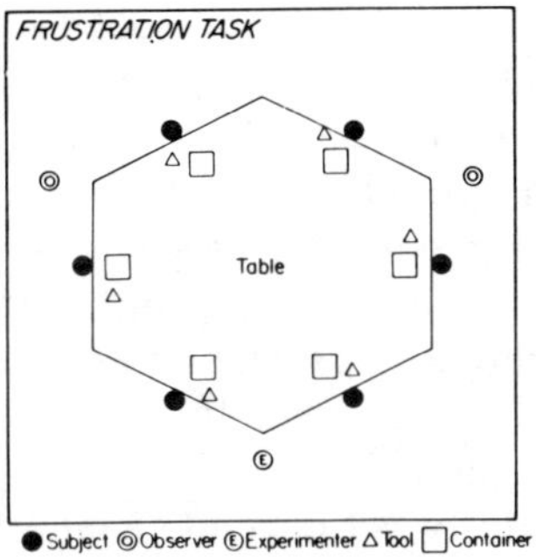

Figure 4-3. Schematic illustration of frustration task.

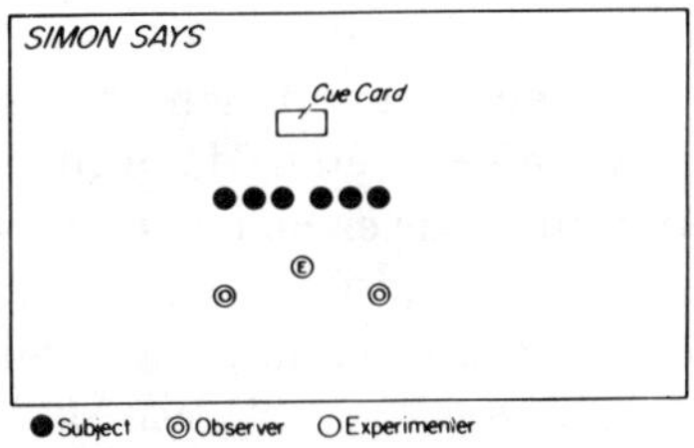

Figure 4-4. Schematic illustration of Simon-says game (impulse control game).

Figures 4-1 through 4-4 from Barrett, D.E. and Radke-Yarrow, M. (1985). Effects of nutritional supplementation on children's responses to novel, frustrating, and competitive situations. *Am. J. Clin. Nutr.* 42: 102–120. Reprinted by permission of American Society for Clinical Nutrition.

First, response to a novel environment (Figure 4–1). Children entered a room with novel toys and play materials. We coded for the latency to first contact with toys, the number of materials touched, and the level of involvement with individual toys. The second was a competitive game (Figure 4–2), which was a balloon-popping relay race. The game was played in teams. Children had to run to the end of the room, pop a balloon with their hands, and run back to touch their teammate. We were interested in the child's capacity to get affectively involved in a physical, exuberant activity and to stay involved with the task. We coded for level of task involvement, positive affect, sense of competition, and fearfulness about popping the balloons. The third situation was a frustrating task (Figure 4–3). Children were seated around a table and given a small, clear plastic container with a prize inside. They were told that if they could get the tops off the containers—the tops were almost impossible to pry off—they could obtain the prize. They were given a small tool, a combination nail clipper-nail file with several flat surfaces, to use if they wished. We obtained data on the number of strategies used to solve the problem. Fourth, there was a Simon Says-type game (Figure 4–4). The experimenter, taking cues from a card behind the children, would say, "One-two-three-yes," or "One-two-three-no." We coded for number of errors of commission (i.e., claps on a "No"), errors of omission, and anticipatory responses. To obtain information on observer agreement, observers watched three children per activity. Thus, there were matching observational records. Coefficient kappa was computed to determine agreement between pairs of raters, correcting for chance agreement. Weighting was linear. Weighted kappas ranged from 0.37 (fearfulness about popping balloons) to 0.95 (sense of competition) with a median of 0.84.

One other variable was scored, and this was the child's manner of participation in three different situations: the new situation and two clean-up tasks. Children were scored on a 1 to 3 scale, with a high score indicating that the child took the initiative in participating (i.e., as a group leader or on his/her own), a low score indicating no participation, and a middle score indicating participation but only as a member of the group. The child's score was his/her average score in the three situations.

At the end of the session, the observers rated children on eight dimensions of behavior. Ratings were on five-point scales. These dimensions were: energy level, work orientation, attention, mood (positive mood = high rating), communication ability, obedience versus disobedience (obedience = high rating), anger, and self-confidence. It should be noted that in determining observer agreement, there was no need to schedule special observations or to attempt to overlap ratings, since each observer rated each child. The child's score on each variable was the average rating across observers.

4.3.4.2 Cognitive testing In addition to measures of social and emotional functioning, cognitive tests were administered. After each of the two morning sessions and after a break for lunch (there also were snack breaks during each morning session), three children from the group came back to the testing center for individual testing. Cognitive tests, standardized by INCAP for use in the villages, were Verbal Inferences, Memory for Objects, Digit Span, Concept Identification, Hidden Figures, and Block Design. Also administered were the Bender-Gestalt Test, a persistence task (persistence on a difficult puzzle), and a test for the ability to follow directions after a period of delay. Those tests assumed to measure attentional capacities were Digit Span, Hidden Figures, the test of persistence, and the task of following directions. In our analysis we also included scores on a cognitive composite consisting of the average of the child's standard scores on the INCAP tests.

4.3.4.3 Reliability of assessments Intraclass correlations for the time-sampling assessments are shown in Table 4–1. Median intraclass correlations for each of the five major categories were 0.81 for interaction with peers, 0.59 for behavior with adults, 0.84 for interaction with the environment, 0.71 for activity level, and 0.66 for emotion. Median values of weighted kappa were 0.40 for the global ratings and 0.84 for responses to structured situations.

Reliability estimates for cognitive and persistence measures were made in the INCAP Longitudinal Study. Test-retest correlation coefficients or coefficients of internal consistency were obtained: Verbal Inferences = 0.87 (test-retest); Persistence = 0.97 (test-retest); Hidden Figures = 0.81 (test-retest); Digit Span = 0.65 (test-retest); Memory for Objects = 0.69 and 0.72 (internal consistency for males and females, respectively); Concept Identification = 0.72 and 0.45 (internal consistency for males and females); Block Design = 0.83 and 0.54 (internal consistency for males and females).

4.3.4.4 Construct validation of the situational measures Before describing the results of the analyses relating nutritional supplementation to behavior, the question of how the structured situation measures of the child's social and emotional characteristics related to natural play measures and observers' ratings is considered. In developing the procedures of this study, a methodological objective was to construct standard situations which would represent challenges of the kind comprising usual life experiences. The abilities to venture into new situations, become positively involved, rise to competitive challenges, cope with frustrations, and control impulses are the characteristics of a competent person. They are also core components of successful social interchange. Therefore, we examined the relation between these aspects of children's responding and independently measured behavior in free-play peer interaction.

Our hypothesis regarding relations between behavior in peer interaction and in the stress situations was that adaptive behaviors in the two kinds of circumstances would be positively associated. Specifically, children who responded with interest in the new environment, demonstrated impulse control, showed persistence and imagination in a frustrating task, and demonstrated the capacity for involvement in a competitive game should show constructive and appropriate free play with peers: friendly, happy involvement at a moderate level of activity.

We also expected that scores in the different situations would be significantly intercorrelated, thereby providing a battery of diagnostic indices of social-emotional competence. For example, we expected that children who showed greater interest and involvement in the new situation would show more interest and involvement in the competitive game, better attention in the impulse control situation, and better performance on the frustrating task. Therefore, each score in each of the structured situations was related to scores based on free-play peer interaction and to every other score on the situational measures.

Finally, we expected that performances on the various individual tasks would be correlated with observers' ratings on several behavior dimensions rated at the end of the entire two-day observation, including attention, energy level, work orientation, disposition, and self-confidence.

4.3.4.4.1 Correlations within tasks Table 4–2 shows the relationships between the structured situation variables.

The relations are strong within tasks, particularly for boys. Within task correlations in the new situation range from 0.77 to 0.89 (in absolute values). In the balloon race, correlations are consistently significant for boys, but less consistent for girls with less than one-half of the correlations significant. In the impulse control game, correlations are high for several relations for boys and girls, but these are relations where there is an operational dependency between the measures; specifically, errors of omission and errors of commission both contribute to total error score.

TABLE 4-2

Correlations among variables measured in structured situations

	New situation			Competitive game				Impulse-control				Frustration	
	1 Lat	2 Exp	3 Inv	4 Inv*	5 Exc	6 Comp	7 Fear	8 Com	9 Ant	10 Omit	11 Tot	12 Strat	13 Init*
1		−8.3‖	−.81‖									−.40‖	−.36§
2	.77‖		.89‖	.53‖	.25‡	.38‖	−.21†	−.24‡	.31§	−.27‡	−.36§	.54‖	.34§
3	.88‖	.92‖		.53‖	.23‡	.40‖	−.21†	−.31‡	.24‡	−.28‡	−.39‖	.54‖	.37‖
4		.26‡	.28‡		.28‡	.55‖	−.37§	−.56‖	.20†	−.49‖	−.70‖	.49‖	.21†
5						.32§	−.24‡			−.21†			
6	−.32‡		.25†		.24†		−.64‖	−.26‡		−.29§	−.41‖	.28‡	.33§
7						−.27‡				−.45‖	.40‖	−.24‡	.22†
8		−.32‡	−.30‡	.79‖							.55‖		
9													
10				−.45‖	−.29‡	−.27‡		.24†			.86‖	−.28‡	−.21
11	−.29‡	−.27‡	−.27‡	−.74‖	−.24†			.68‖		.87‖		−.33§	
12	−.39§	.54‖	.44‖	.51‖				−.56‖		−.22†	−.44‖		.28‡
13		.31‡	.32‡							.34§	−.27‡	.26‡	

Note: Correlations for boys are above the diagonal. Correlations for girls are below the diagonal. Abbreviations are: Lat: Latency to respond, Exp: Number of materials touched, Inv: Level of involvement, Exc: Excitement, Comp: Sense of competition, Fear: Fearfulness, Com: Errors of commission, Ant: Anticipates response, Omit: Errors of Omission, Tot: total number of errors, Strat: Number of strategies used, Init: Average level of initiative across four situations. All n's for boys are 78, except for correlations involving the impulse control game, for which n's are 77, and initiative, for which n's are 76. All n's for girls are 60, except for correlations involving the impulse control game, for which n's are 57, and initiative, for which n's are 48.

* In raw score coding for these variables, high scores indicated low involvement or initiative. The directions of correlations have been reversed for the purpose of presentation; ie, high scores indicate high involvement or initiative.

† $p < .10$ ‡ $p < .05$ § $p < .01$ ‖ $p < .001$.

From Barrett, D.E. and Radke-Yarrow, M. (1985) Effects of nutritional supplementation on children's responses to novel, frustrating, and competitive situations. *Am. J. Clin. Nutr.* 42: 102–120. Reprinted by permission of American Society for Clinical Nutrition.

4.3.4.4.2 Correlations across tasks Across tasks there are significant relationships for both sexes. These are also shown in Table 4–2. For boys, responses to the new environment were highly predictive of responses to the frustration task and significantly related also to performance in the impulse control situation and the competitive game. Children who were highly involved in the new environment, touching many toys and exploring materials for longer periods of time, were more involved and less inhibited in the competitive game, quicker and more accurate in the impulse control task, and more able to respond adaptively to the frustrating task than were children who were not highly involved. For girls, involvement in the new situation predicted involvement in the balloon race but less strongly than for boys. Involvement in the new situation, however, strongly predicted the number of strategies used on the frustrating task. Finally, for both sexes, high involvement in the new situation was predictive of level of initiative across the three situations.

For boys, behavior in the competitive game was generally predictive of behavior in the other situations. In particular, the involvement variable, the competition variable, and the fearfulness score predicted performance in the impulse control, frustration and initiative tasks. For girls, there were high correlations between involvement in the competitive game and behavior in the impulse control situation and behavior in the frustrating task, and also with overall level of initiative. Finally, for girls, performance on the impulse control game was predictive of performance in the frustrating situation and level of initiative across situations. These relationships were less strong for boys.

On the whole, the findings support the supposition that the competencies measured in the different situations—the ability to show interest and involvement in a new environment, to participate in a competitive game, to control motor impulses, to respond appropriately to a frustrating problem and to take the initiative in a variety of situations—are closely interrelated.

4.3.4.4.3 Associations between behaviors in structured situations and peer play The next set of analyses concern relations between behaviors in the specific situations and behaviors in the free play situations.

The details of observing and coding free play activities were reported earlier. Briefly, each child was observed for approximately 25 two-minute intervals. Behaviors were coded in five classes: Interactions with peers, contact with adults, involvement with the physical environment, level of activity, and emotion. The categories are shown in Table 4–1.

In the present analyses, our interest was in the following dimensions of behavior: level of involvement with peers and with group activities, vigor of physical activity, and expression of positive and negative emotions.

Again, findings are persuasive; results are shown in Tables 4–3 and 4–4. For both boys and girls, behaviors in the stressful situations were highly related to friendly interaction, activity level, affect, and involvement in group activities in natural play situations.

The major findings may be summarized as follows: Children who showed a moderate activity level were highly interested in and involved in the new situation, were very involved in the balloon race, appropriately responsive and nonimpulsive in the Simon Says game, and persistent on the frustrating problem. High activity level was correlated negatively with the same variables. An interesting finding was that for boys, very high activity level (as would occur if the child were involved in a very physical game during free play) was correlated positively with performance in all situations. Thus, unlike high activity level (i.e., which would indicate nervous activity) very high activity level indicated degree of involvement in active play. There were fewer correlates with very high activity level for girls, but they were in the same direction as for boys.

Findings for happy affect were extremely consistent. For both boys and girls, happy affect was related positively to capacity for interest and involvement in the new situation, involvement in the balloon game, impulse control, persistence on the frustrating task, and level of participation across tasks. Correlations with anxiety were in the opposite direction. Correlations between anxiety in free play situations and low involvement in the competitive game and errors of impulse control were high, ranging from 0.60 to 0.78. There were significant correlations between amount of routine friendly interaction and involvement in group activities and behavior in the structured situations. The correlations were parti-

TABLE 4-3

Correlations between structured situation variables and free play variables: boys

	Friend	Phys	Nobeh	Involv	Low	Mod	High	V High	Happy	Sad	Angry	Anxious	No emot
New situtіon													
Latency								−.44§					
No. Items Touched				.42§	.20*	.33‡	−.23*	.40§	.30‡		.22†	−.38§	
Involvement				.43§		.38§	−.30‡	.42§	.31‡			−.41§	
Balloon Race													
Involvement	.29§		−.33‡	.81§		.57§	−.55§	.26†	.63§			−.70§	−.30‡
Excitement	.27‡	.25†	−.30‡		−.26†				.28†				−.27†
Competition	.45§		−.47§	.47§	−.29‡	.30‡		.26†	.36‡				−.32‡
Fear	−.44§		.46§	−.29‡	.33‡	−.28†		−.21*	−.39§				.41§
Simon Says													
Committed				−.58§	−.27†	−.42§	.53§		−.31‡				
Anticipated								.21*			.40§		
Omitted	−.21*		.24†	−.41§		−.40§	.21*	−.21*	−.49§			.55§	.22*
Total Errors	−.22*		.24†	−.64§		−.50§	.42§	−.22*	−.56§			.73§	.19*
Frustration													
Strategies	.19*		−.22*	.41§		.30‡	−.30‡	.25†	.33‡			−.34‡	
Initiative	.40§		−.42§	.20*	−.40§		.23*	.23†	.25†		.21*		−.32‡

Note: Abbreviations for free play variables are *Friend:* Friendly Interaction; *Phys:* Physical aggression; *Nobeh:* No interaction with peers; *Involv:* Involved in the Group activity; *Low:* Low activity level; *Mod:* Moderate activity level; *High:* High activity level; *V High:* Very high activity level; *Happy:* Happy affect; *Sad:* Sad affect; *Angry:* Angry affect; *Anx:* Anxious expression; *No Emot:* No emotion shown.

* $p < .10$ † $p < .05$ ‡ $p < .01$ § $p < .001$.

From Barrett and Radke-Yarrow, 1985.

TABLE 4-4

Correlations between structured situation variables and free play variables: girls

	Friend	Phys	Nobeh	Involv	Low	Mod	High	V High	Happy	Sad	Angry	Anxious	No emot
New situtlon													
Latency	$-.32^{*}$		$.31^{*}$		$.23^{\dagger}$		$-.28^{\dagger}$	$-.48^{\ddagger}$	$-.29^{*}$				$.23^{\dagger}$
No. Items Touched				$.43^{\S}$		$.25^{*}$		$.27^{*}$	$.38^{\ddagger}$				$-.23^{\dagger}$
Involvement				$.42^{\S}$	$-.22^{\dagger}$	$.28^{*}$		$.31^{*}$	$.38^{\ddagger}$				$-.24^{\dagger}$
Balloon Race													
Involvement	$-.33^{\ddagger}$		$-.64^{\S}$	$.67^{\S}$		$.60^{\S}$	$-.51^{\S}$	$-.32^{\ddagger}$	$.35^{\ddagger}$			$-.60^{\S}$	
Excitement	$.40^{\S}$		$-.35^{\dagger}$		$-.28^{*}$	$.61^{\S}$	$-.22^{\dagger}$		$.25^{*}$				$-.39^{\dagger}$
Competition		$-.28^{\dagger}$							$.23^{\dagger}$		$-.39^{\ddagger}$		$-.39^{\dagger}$
Fear													
Simon Says													
Committed	$.28^{*}$		$-.42^{\S}$	$-.66^{\S}$		$-.64^{\S}$	$+.63^{\S}$		$-.50^{\S}$			$.78^{\S}$	$-.27^{*}$
Anticipated													
Omitted				$-.38^{\ddagger}$		$-.40^{\S}$			$-.33^{*}$			$.25^{\dagger}$	
Total Errors		$.66^{\S}$	$-.37^{\ddagger}$	$-.61^{\S}$		$-.63^{\S}$	$.46^{\S}$		$-.50^{\S}$			$.59^{\S}$	
Frustration													
Strategies			$-.23^{\dagger}$	$.70^{\S}$		$.60^{\S}$	$-.33^{*}$		$.45^{\S}$	$-.35^{\ddagger}$		$-.49^{\S}$	
Initiative	$.48^{\S}$		$-.30^{\ddagger}$	$.37^{\ddagger}$	$-.53^{\S}$	$.27^{*}$		$.25^{\dagger}$	$.28^{*}$	$-.34^{\ddagger}$			

Note: Abbreviations for free play variables are *Friend:* Friendly Interaction; *Phys:* Physical aggression; *Nobeh:* No interaction with peers; *Involv:* Involved in the Group activity; *Low:* Low activity level; *Mod:* Moderate activity level; *High:* High activity level; *V High:* Very high activity level; *Happy:* Happy affect; *Sad:* Sad affect; *Angry:* Angry affect; *Anx:* Anxious expression; *No Emot:* No emotion shown.

* $p < .10$ † $p < .05$ ‡ $p < .01$ § $p < .001$.

From Barrett and Radke-Yarrow, 1985.

cularly strong for boys. Involvement in the group activity was a strong predictor of interest and involvement in the new environment; involvement in the balloon game ($r = 0.81$); competition, excitement, and lack of fearfulness in the balloon game; and accurate responding in the impulse control situation. In addition, high involvement was predictive of level of initiative across situations. Similar correlations were observed for frequency of friendly interactions with peers. The only difference was that behavior in the new situation was not related to amount of friendly interaction, while involvement in the balloon race, performance in the impulse control game, and level of participation across tasks were.

For girls, patterns of correlations were similar. The highest correlation ($r = 0.70$) was between level of involvement in group situations and number of strategies used in the frustrating task.

4.3.4.4.4 Associations between behaviors in structured situations and global ratings At the end of each two-day session, observers and the experimenter rated each child on a one-to-five scale for level of attention, energy level, work-orientation, positive disposition, and confidence. Ratings also were made for anger, noncompliance (disobedience), and communication ability.

Global ratings and their relationships to the situation variables are shown in Table 4–5. There were consistent correlations between behaviors in the structured situations and the five ratings. For boys, involvement in the balloon race, involvement in the Simon Says game, and level of initiative were positively related to observer ratings of energy level, work orientation, mood, level of communication, and self-confidence. Self-confidence also was related to interest and involvement in the new situation. It is important to note that operationally, these variables were not independent. Since the ratings were made after both half-day sessions, observers were free to make judgments about children based on known performances in various situations. These finding suggest that for boys, behavior in the structured situations either influenced the observers' ratings or was related to other behaviors (i.e., free play behaviors) which did. For girls, the only rating variables with consistent correlates were mood and self-confidence. Mood was positively related to involvement in the new situation and initiative; confidence was positively related to behavior in the

TABLE 4-5

Correlations between structured situation variables and global ratings

	Energy		Work		Anger		Mood		Attention		Common		Confi		Disob	
	B	G	B	G	B	G	B	G	B	G	B	G	B	G	B	G
New Situation																
Latency		−.23†						−.37‡		−.28*						
No. Items Touched								.26†	.24*		.21†		.31‡	.29†		
Involvement			.23*					.23†	.19†		.28†		.22*	.23†		
Balloon Race																
Involvement	.25*		.23*				.26*	+.28†			.28‡		.30‡			
Excitement	.39§		.21†		.21†		.30‡				.51§		.39§	.26†	.34‡	
Competition	.27*		.27*				.31‡				.34‡		.36§			
Fear	−.29‡	−.24†	−.32‡				−.30‡		−.24*		−.37‡		−.41§			
Simon Says																
Committed Anticipated											.23*		.22†			
Omitted	−.32‡		−.28*				−.19†				−.32‡		−.32‡	−.28*		
Total Errors	−.26*		−.24*								−.28*		−.28*	−.29*		
Frustration																
Strategies			−.28*	.24†					.22†		.20†		.24*	.31*		
Initiative	.44§	+.35‡	.38§	.55§			.45§	.35‡	.21†	+.55§	.34‡	.44§	.46§	.55§		

* p < .05 † p < .10 ‡ p < .01 § p < .001

From Barrett and Radke-Yarrow, 1985.

new situation, the impulse control situation, the frustration task, and, most strongly, level of initiative across tasks.

These findings strongly support the construct validity of the situation measures. They suggest that the variables are measuring the outcomes in which we are interested.

The fact that our predictions about the interrelationships among behaviors were born out not only increases our confidence in the validity of our measures of children's social competencies, it helps to provide an integrative framework for thinking about children's prosocial responding, in particular in relation to malnutrition.

An important result of our analyses was our identification of networks of relationships among variables that we felt were related as measures of the child's adaptive and prosocial competencies. The relationships are illustrated in Figures 4–5 to 4–10. Figure 4–5 illustrates the relationships among the structured situation variables. Figures 4–6 to 4–10 show how the structured

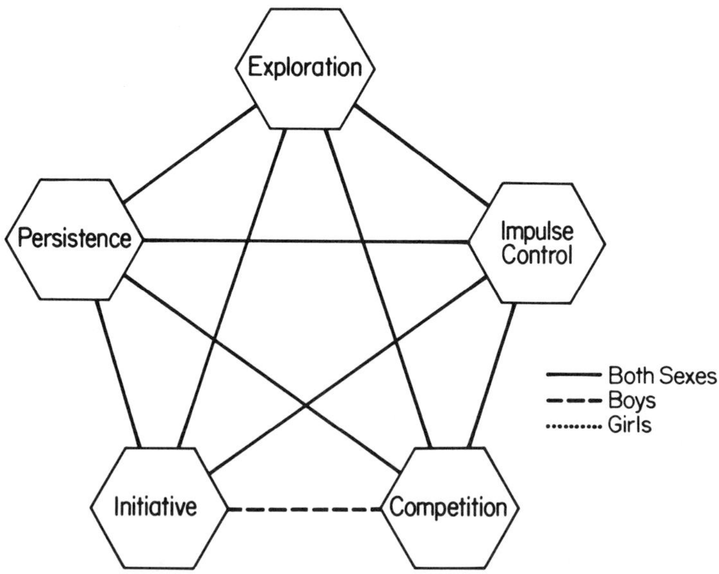

Figure 4-5. Associations between children's responses to routinely stressful or challenging situations.

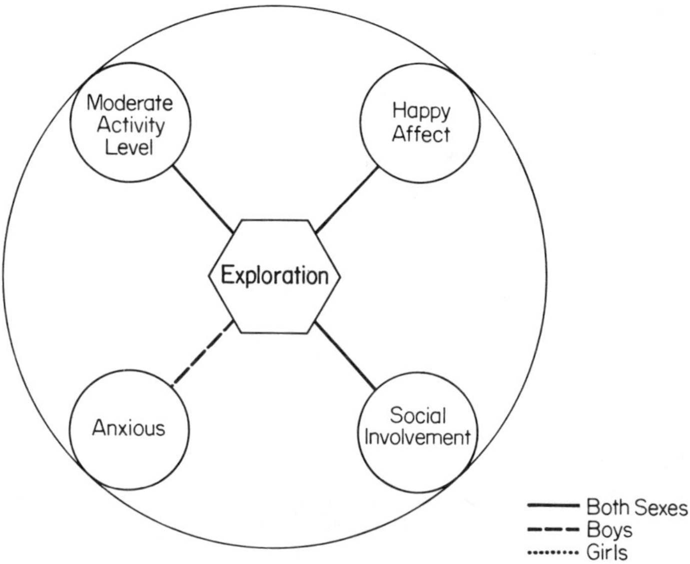

Figure 4-6. Associations between exploration in a novel situation and behaviors in free-play activities.

Figures 4-5 and 4-6 from Barrett, D.E. (1986). Behavior as an outcome in nutrition research. *Nutr. Rev. Supp.* 44: 224–236. Reprinted by permission of International Life Sciences Institute-Nutrition Foundation.

situation variables relate to four aspects of behavior in natural situations, all of which were significantly intercorrelated: happy affect, moderate activity level, social involvement, and low anxiety.

In the figures, lines drawn between situations or categories indicate that for at least one pair of variables across those situations, there was a correlation significant at $p < .10$. (See Tables 4–2 to 4–5 for the complete sets of correlations.) The figures help us to recognize the relationship of the specific measures of social and emotional functioning that we examined in relation to malnutrition to the child's overall well-being.

We see from our analyses a complex of behaviors related to undernutrition—behaviors which provide a conceptualization of healthy functioning in the school-age child. The abilities to be-

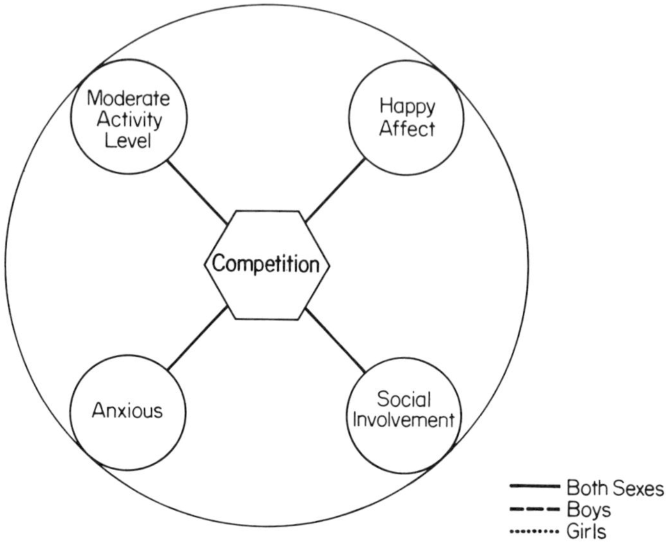

Figure 4-7. Associations between involvement in a competitive game and behaviors in free-play activities.

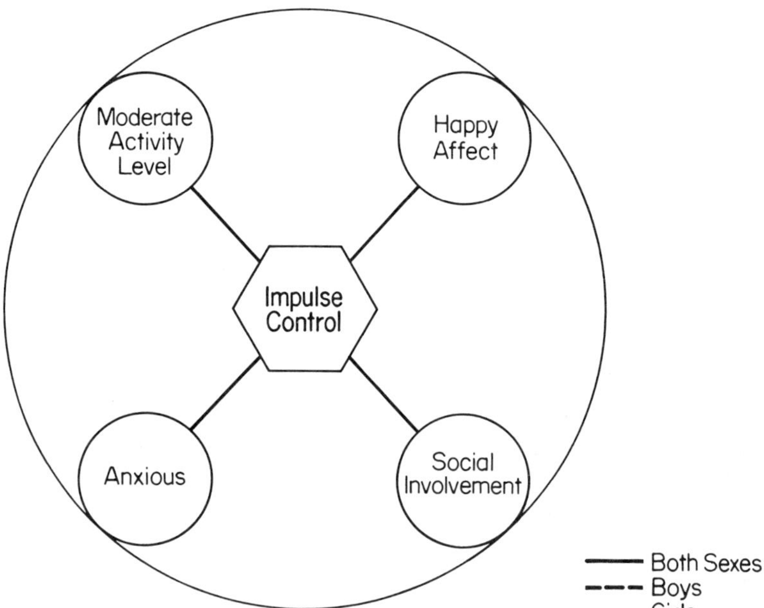

FIgure 4-8. Associations between impulse control in a Simon-says type game and behaviors in free-play activities.

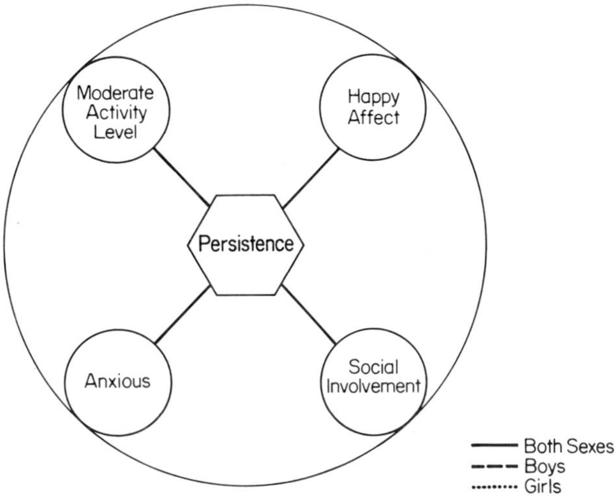

Figure 4-9. Associations between persistence on a frustrating problem and behaviors in free-play activities.

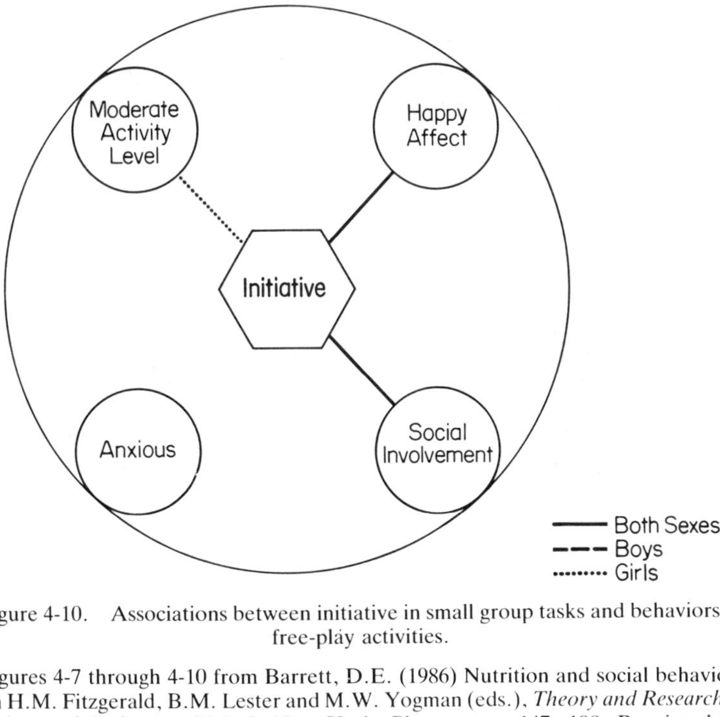

Figure 4-10. Associations between initiative in small group tasks and behaviors in free-play activities.

Figures 4-7 through 4-10 from Barrett, D.E. (1986) Nutrition and social behavior. In H.M. Fitzgerald, B.M. Lester and M.W. Yogman (eds.), *Theory and Research in Behavioral Pediatrics*, Vol. 3. New York: Plenum, pp. 147–198. Reprinted by permission of the publisher.

come involved with other children, to enjoy interaction, to be moderately assertive (i.e., not timid), and to regulate one's motor activity are characteristics of the healthy child. We see, in addition, the abilities to deal with routinely stressful problems of the kind encountered in everyday life: responding to challenging situations, venturing into new areas, persisting on difficult problems, controlling impulses, and taking the initiative in ambiguous situations.

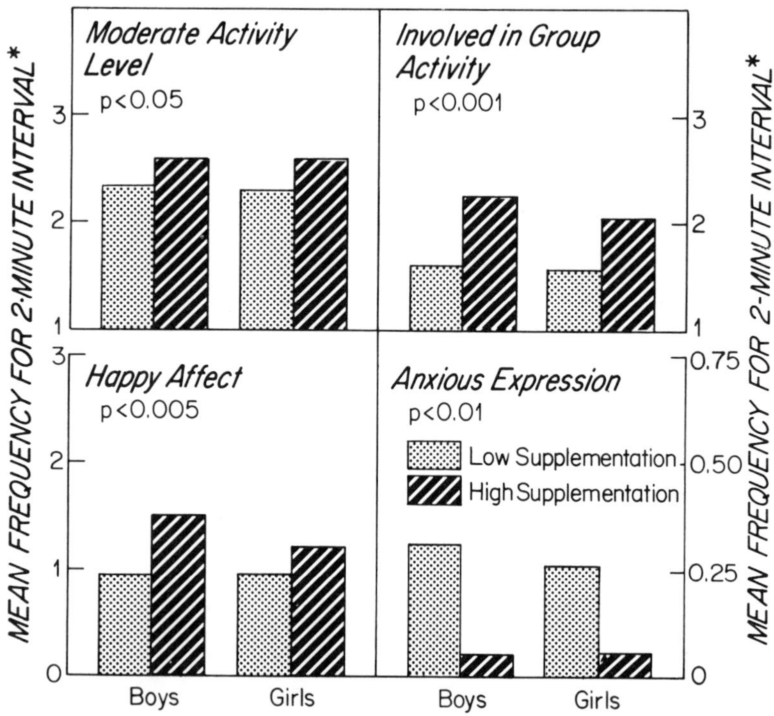

*Each 2-minute interval was divided into 4 30-second intervals. Minimum Score = 0. Maximum Score = 4.

Figure 4-11. Variables for which effects of supplementation are consistent across sexes.

From Barrett and Radke-Yarrow, 1985.

4.4 *Results*

4.4.1 *Results of the multiple regression analyses*

Results of the multiple regression analyses for the prediction of social-emotional outcomes across villages are presented in detail in Table 4–6. Table 4–7 summarizes results of multiple regression analyses across and within villages. In the tables, dependent

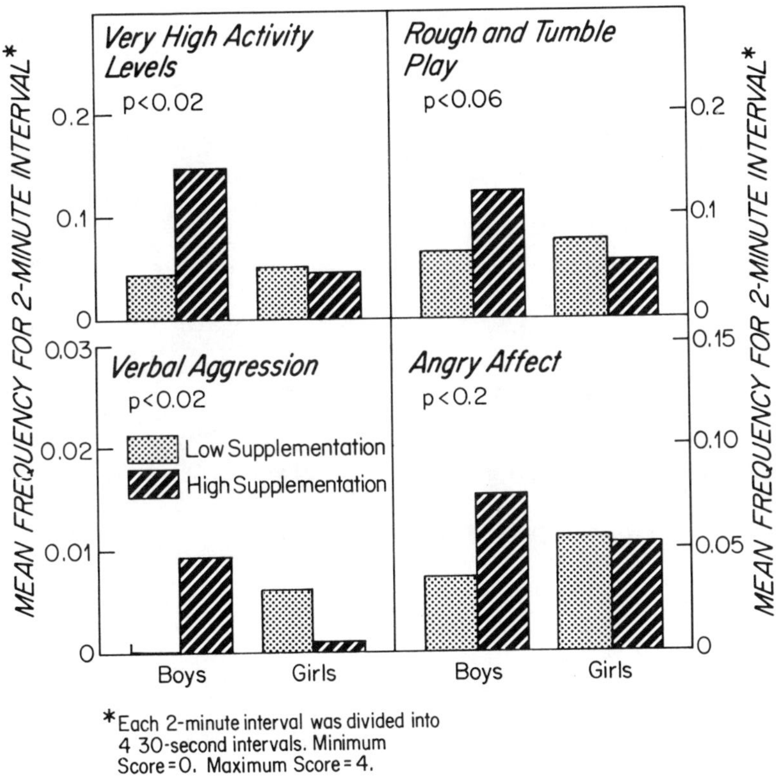

Figure 4-12. Variables for which effects of supplementation differ for boys and girls.

From Barrett and Radke-Yarrow, 1985.

TABLE 4-6

Multiple regression analysis for prediction of social–emotional variables

Dependent variables	Step 1	Step 2				Step 3					Step 4					
	r	R	β_{SES}	β_{MS}	F_{MS}	R	β_{SES}	β_{MS}	β_{CS1}	F_{CS1}	R	β_{SES}	β_{MS}	β_{CS1}	β_{CS2}	F_{CS2}
1. Helps	.06	.07	−.06	.01	.01	.18	−.03	.12	−.21*	3.64*						
2. Seeks help from peer	.07	.25**	.10	.24***	7.45***	.29**	.07	.14	.18*	2.89*						
3. Defends self	.06	.12	.07	.11	1.50	.13	.07	.09	.04	.10	.22	.09	.07	−.17	.28**	4.12**
4. Negative to bid from peer	.08	.08	.07	−.03	.14	.17	.10	.07	−.18*	2.79*						
5. No interaction with adult	−.09	.10	−.08	.05	.31	.18	−.12	−.05	.18*	2.81*						
6. Involved in group activity	.12	.20*	.15	.16*	3.03*	.27**	.10	.03	.23**	4.66**						
7. Only watching	−.21**	.32***	−.24**	−.25**	6.57**											
8. No activity	−.22**	.27***	−.24***	−.17*	3.53*											
9. Moderate activity level	.07	.13	.09	.11	1.53	.24*	.00	−.02	.24**	4.82**						
10. Very high activity level	.09	.10	.09	.06	.37	.19	.06	−.05	.19*	3.11*						
11. Happy	.15*	.19*	.17*	.13	1.94	.29**	.12	−.02	.26**	5.88**						
12. Angry	.04	.10	.05	.10	1.10	.24*	.01	−.05	.27**	6.19**						
13. Anxious	−.12	.15	−.13	−.09	.90	.27**	−.08	.07	−.28**	6.74**						

Note: SES = socioeconomic status; MS = mother supplementation; CS_1 = child supplementation from birth to 2 years; CS_2 = child supplementation, 2–4 years. Predictor variables for the steps are as follows: Step 1, SES; Step 2, SES and MS; Step 3, SES, MS, CS_1; Step 4, SES, MS, CS_1, CS_2. Significance levels for r indicate significance of regression equation with 1, 122 degrees of freedom; for R, $df = k, 124 - (k - 1)$. k is the number of variables in the equation. β values show standardized regression coefficients with all variables in the equation at a given step in the analysis. F values represent increase in R^2 associated with the entry of the last variable in the equation. Results are reported up to and including the final step in the equation where a last-entered variable makes a significant increase in R^2.

* $p < .10$. ** $p < .05$. *** $p < .01$.

From Barrett, Radke-Yarrow and Klein, 1982.

variables significantly associated with one or more independent variables are presented. Because of the clarity of the theoretical framework in which we have hypothesized empirical relations, results are reported for significance levels at or below 0.10. Attendance to marginally significant associations helps to trace out the relationships within the hypothesized network of variables.

4.4.1.1 Social-emotional measures Table 4–6 gives results for the multiple regression analysis for the prediction of social and emotional characteristics across villages. For each dependent variable the table shows the simple correlation of the variable with the SES composite and, at each succeeding step of the analysis, the resulting multiple correlations, beta weights, and *F* values for the increment in the squared multiple correlation due to the entry of the last-entered variable. The results are reported up to and including the last independent variable that significantly increased the value of the squared multiple correlation.

Three variables were significantly predicted by SES alone. There were negative correlations between SES and only watching other children, and between SES and no activity (sum of only watching and inactive), and a positive correlation with happy affect. The first supplementation variable representing mother supplementation during pregnancy (MS) significantly increased the variance accounted for in the first two of these measures. Better maternally supplemented children (controlling for SES) were less often only watching and less often showed no activity than did poorly supplemented children. Children of better-supplemented mothers more often requested help from a peer.

The addition of the variable representing child supplementation from birth to 2 years (CS1) significantly increased the multiple correlation for ten variables. High caloric supplementation during this period was positively associated with Seeks Help from Peer, Involved in Group Activity, Moderate Activity Level, Happy Affect, and Angry Expression (better-supplemented children showed anger more often). High supplementation was negatively related to anxious expression, and to negative response to bid for help from peer. There also was a marginal negative relationship with Helps Peer, and a marginal positive association with Very

TABLE 4-7

Summary of multiple regression analyses

Predictor variable	Across villages SES	MS	CS1	CS2	Calorie-only village SES	MS	CS1	CS2	Protein-calorie village SES	MS	CS1	CS2
Friendly	.10	.12	−.00	−.08	.02	.19	.05	.50**	.23	.20	.09	−.25
Helps	−.07	.01	−.21*	.00	−.19	.22[a]	−.30**	.06	.15	−.18	−.38*	.11
Intrudes	.11	.01	.08	−.10	.21	−.15	.09	.29	.15	.10	.40*	−.10
Seeks help	.07	.24**	.18*	.06	−.02	−.00	.16	.26	.12	.43***	.03	.03
Defends self	.06	.11	.04	.28**	.20	.14	.21	.41**	.05	.14	.15	.14
Negative to bid	.08	−.03	−.18**	.02	.16	.02	−.07	−.01	.01	.11	.19	.08
No interaction (peer)	−.13	−.10	.06	.06	−.12	−.18	.02	−.41**	−.22	−.22	−.11	.26
Seeks contact with adult	.09	.05	−.12	.07	.15	.15	.33**	−.04	.00[b]	.00[b]	.00[b]	.00[b]
No interaction (adult)	−.09	.05	.18*	.07	−.08	.05	.14	.19	−.12	−.08	−.18	.04
Involved in group activity	.12	.16*	.23**	.04	.08	.12	.10	.22	.18	.30**	−.31	−.23
Plays with materials	.15*	.06	.05	.17	.19	−.02	.04	.44**	.22	.09	.17	.15
Imaginative play	−.05	−.04	−.08	.08	.07	−.02	.08	.36*	−.20	−.03	−.28	−.22
Only watching	−.21**	−.25**	.02	.08	−.07	−.39**	.23	−.21	−.33**	−.31**	.09	.16
No activity (environment)	−.22**	−.17*	−.01	−.08	−.17	−.11	.02	−.55***	−.32**	−.34**	.13	.09
Low activity level	−.09	−.12	.04	.02	.06	−.18	−.02	−.35*	−.32**	−.23	.03	.13
Moderate activity level	.09	.11	.24**	−.02	−.02	.06	.17	.20	.13	.11	.06	−.15
Very high activity level	.15*	.06	.19*	.06	.31**	.10	−.01	.14	.12	.18	−.16	−.29
Happy	−.12	.13	.26**	−.00	.11	.15	.36**	.32*	.23*	.17	.16	−.22
Anxious	.04	−.09	−.28**	−.06	−.12	.00	−.26*	−.10	−.19	−.06	.02	.20
									.06			

Angry		.10	.27**	.17	.08	.05	.07	.23		.02	.37*	.22
Persistence	.03	.06	.01	.32**	.13	.06	−.10	.44**	.07	.12	.11	.40**
Distracted	−.18**	−.14	−.01	−.09	−.26*	−.13	−.12	−.10	−.09	−.18	.06	.01
Following Directions	.00	−.02	.02	.34**	−.01	.21	.04	.30	−.16	−.06	.13	.35*
Hidden Figures	.11	.17*	−.13	−.02	.29**	.36***	−.26*	.44**	−.07	.13	.28	−.03
Digit Span	.19**	.02	.02	.14	.15	.06	.07	.16	.16	−.05	−.04	.12
Block Design	.16*	.03	.03	−.13	.22*	.01	−.03	.41**	.13	.04	.14	−.17
Concept Identification	.23**	−.00	−.04	.00	.19	.11	.09	.21	.26*	−.18	.20	−.02
Bender Gestalt[c]	−.13	−.08	.06	−.07	−.16	−.14	−.00	−.34*	−.11	−.08	.11	−.04
Cognitive Composite	.19**	.06	.01	.08	.22*	.15	−.11	.44**	−.19	−.03	.23	.04

Note: Entries in the table are standardized partial regression coefficients (beta weights). Each shows the weight given a particular independent variable in predicting the criterion at the step at which it was entered into the equation. Significance levels indicate the significance of the increase in R^2 due to the entry of the independent variable. SES = socioeconomic status; MS = mother supplementation; CS1 = child supplementation from birth to 2 years; CS2 = child supplementation, 2–4 years. [a] MS had the strongest association with helps at Step 3 (β = .36, p < .05). CS1 acted as a supressor variable. [b] Regression statistics were not computed due to insufficient variability in the criterion variable. [c] High scores on the Bender Gestalt indicate poor performance.

* p < .10. ** p < .05. *** p < .01.

From Barrett, Radke-Yarrow and Klein, 1982.

High Activity Level. Better-supplemented children interacted less frequently with an adult observer.

The fourth variable in the equation, representing child supplementation from 2 to 4 years (CS2) significantly increased the prediction of only one measure—Defends Self—and assumed a positive beta weight in the equation.

Considered together, the findings indicate that maternal caloric supplementation during pregnancy and child caloric supplementation during the first two years were related to high levels of social responsiveness, interaction with the environment, activity level, and expression of affect—both positive (happy) and negative (angry)—at school age.

4.4.1.2 Individual tests As shown in Table 4–7, SES was positively correlated with Digit Span (r = 19; $p < .05$), Concept Identification ($r = .23$, $p < .05$), Block Design ($r = .16$, $p < .10$), and Cognitive Composite ($r = .19$, $p < .05$). SES was negatively related to amount of time distracted by attractive toys during work on the impossible puzzle (r = *.18*, $p < .05$). None of the supplementation variables significantly improved the prediction of these measures. MS significantly improved prediction of Hidden Figures performance, $F(1,121) = 3.56$, $p < .10$, with better maternally supplemented children scoring higher. CS1 did not significantly improve prediction of any measures. CS2 improved the prediction of Persistence on the Impossible Puzzle, $F(1,119) = 5.29$, $p < .05$, and performance on the Following Directions test, $F(1,119) = 4.92$, $p < .05$. The variable CS2 was weighted positively in both analyses.

In summary, SES but not the supplementation variables predicted overall cognitive performance; higher SES children performed better than lower SES children. Of the supplementation measures, only one—supplementation from 2 to 4 years—significantly improved the prediction of more than one dependent measure, and this occurred on tests that tapped primarily attentional characteristics (i.e., Persistence and Following Directions).

4.4.2 Results of the analyses of variance

Results of the analysis of variance for all social-emotional measures are summarized in Tables 4–8, 4–9, and 4–10.

TABLE 4-8

Results of analysis of variance for effects of supplementation, sex, and supplementation × sex interaction on children's behavior: structured situations

Variable	Means for boys		Means for girls		Significant effects
	Low supp	High supp	Low supp	High supp	
Novel Environment					
No. of Materials Touched*	1.45 (.50)	1.82 (.62)	1.40 (.61)	1.57 (.74)	Supp[‖]
Level of Exploration[†]	1.83 (.95)	2.29 (.90)	1.53 (.80)	1.75 (.92)	Supp[‖], Sex[¶]
Competitive Game					
Level of Involvement[†‡]	1.78 (.87)	1.14 (.47)	1.59 (.87)	1.28 (.65)	Supp[¶]
Frustrating Task					
No. of Strategies Used*	1.97 (1.09)	2.56 (.97)	2.15 (1.46)	2.64 (1.06)	Supp[¶]
Impulse Control Game					
Anticipates Response*	.038 (.196)	.400 (.708)	.000 (.000)	.000 (.000)	Supp. Sex[¶] Int[§]
Error of Commission*	.730 (1.00)	.900 (.900)	.434 (.589)	.400 (.707)	Sex[¶]
Error of Omission*	1.23 (2.74)	.32 (1.43)	1.00 (2.33)	.60 (1.50)	Supp[§]

Note: Supplementation effects are indicated by Supp, Interaction effects by Int., Sex effects by Sex. Significance levels are based on F tests with 1,134 df (for Novel Environment, Competitive Game, and Frustrating Task) or 1,110 df (Impulse Control Game). F values are given in the text. Standard deviations are given in parentheses.

* Scores represent raw frequencies.

[†] Scores are based on a rating scale.

[‡] Low score indicates high involvement.

[§] $p < .10$.

[‖] $p < .05$.

[¶] $p < .01$.

From Barrett and Radke-Yarrow, 1985.

4.4.2.1 Supplementation effects Main effects for supplementation occurred for several of the behaviors measured on the Interaction Record. High-supplemented children had higher rates of behavior than low-supplemented children on Involved in Group

TABLE 4-9

Results of analysis of variance for effects of supplementation, sex, and supplementation × sex interaction on children's behavior: time-sampling assessments

Variable	Means for boys		Means for girls		Significant effects
	Low supp	High supp	Low supp	High supp	
Verbal Aggression	.000 (.000)	.009 (.018)	.005 (.018)	.001 (.006)	Int†
Rough and Tumble Play	.063 (.098)	.122 (.142)	.073 (.119)	.054 (.093)	Int*
Seeks Help from Peer	.183 (.197)	.233 (.214)	.110 (.156)	.160 (.173)	Sex†
Negative to Bid from Peer	.022 (.038)	.020 (.031)	.206 (.585)	.163 (.554)	Sex†
Involved in Group Activity	1.61 (1.19)	2.26 (.707)	1.59 (.943)	2.06 (.640)	Supp‡
Talks to Adult	.035 (.066)	.058 (.076)	.088 (.115)	.057 (.082)	Sex*, Int*
Wanders	.012 (.031)	.008 (.021)	.040 (.091)	.012 (.028)	Supp*, Sex*
Distracted	.117 (.103)	.117 (.135)	.199 (.196)	.108 (.094)	Supp*, Int*
Moderate Activity Level	2.33 (.909)	2.59 (.541)	2.29 (.987)	2.61 (.660)	Supp†
Very High Activity Level	.042 (.116)	.148 (.177)	.050 (.092)	.047 (.077)	Supp‡, Sex†, Int†
Happy	.948 (.801)	1.48 (.748)	.924 (.649)	1.18 (.528)	Supp‡
Anxious	.301 (.579)	.048 (.195)	.262 (.647)	.062 (.330)	Supp‡

Note: Supplementation effects are indicated by Supp, Interaction effects by Int., Sex effects by Sex. Significance levels are based on F tests with 1,134 df except for the variables Verbal Aggression and Distracted for which df are 1,110. F values are given in the text. Scores are in units of frequency per 2-min interval. Standard deviations are given in parentheses.

* p < .10.
† .05.
‡ p < .01.

From Barrett and Radke-Yarrow, 1985.

TABLE 4-10

Results of analysis of variance for effects of supplementation, sex, and supplementation × sex interaction on children's behavior: global ratings

Variable	Means for boys		Means for girls		Significant effects
	Low supp	High supp	Low supp	High supp	
Energy Level	2.71 (.78)	3.41 (.94)	2.82 (.83)	2.96 (.87)	Supp§, Int†
Anger	1.12 (.32)	1.34 (.72)	1.07 (.26)	1.19 (.49)	Supp†
Good Disposition	2.82 (.82)	3.41 (.89)	3.03 (.92)	2.94 (.80)	Int‡
Compliance*	1.10 (.65)	1.63 (1.42)	1.10 (.56)	1.32 (1.10)	Supp‡
Self-Confidence	2.79 (.99)	3.24 (.96)	3.00 (.95)	3.17 (.83)	Supp†

Note: Supplementation effects are indicated by Supp., Interaction effects by Int., Sex effects by Sex. Significance levels are based on F tests with 1,128 df. F values are given in the text. Scores are based on five point rating scales (1−5). Standard deviations are given in parentheses.

* High score indicates low compliance.

† $p < .10$

‡ $p < .05$

§ $p < .01$

From Barrett and Radke-Yarrow, 1985.

Activity, $F(1,134) = 4.53$, $p < .001$; Moderate Activity Level, $F(1,134) = 5.59$, $p < .04$; Very High Activity Level, $F(1,134) = 5.59$, $p < .02$; and Happy Affect, $F(1,134) = 10.86$, $p < .001$. High-supplemented children had lower rates of behavior on Anxious, $F(1,134) = 7.90$, $p < .01$; Distracted, $F(1,134) = 3.02$, $p < .09$; and Wanders, $F(1,134) = 3.77$, $p < .06$. On global ratings, better-supplemented children were rated as more Energetic, $F(1,128) = 7.48$, $p < .01$, and more Self-confident, $F(1,128) = 3.40$, $p < .07$. They had higher ratings on Anger, $F(1,128) = 3.61$, $p < .06$, and lower ratings on Compliance, $F(1,128) = 4.03$, $p < .04$.

With respect to responses to specific situations, high-supplemented children had higher scores than low-supplemented children on Number of Materials Touched, $F(1,134) = 6.28$, $p < .02$, and Level of Involvement with Toys in the Novel Environment, $F(1,134) = 4.72$, $p < .04$; scored higher on Level of Task Involvement in the Competitive Game, $F(1,134) = 13.89$, $p < .001$, and used a Larger Number of Strategies, $F(1,138) = 7.52$, $p < .01$ on the Frustrating Problem. In the Impulse Control Game, they made more Anticipatory Responses, $F(1,110) = 4.76$, $p < .05$, and made fewer Errors of Omission, $F(1,110) = 2.88$, $p < .10$.

There were no supplementation effects with respect to performance on cognitive tasks.

4.4.2.2 Sex effects For the Interaction Record measures, boys had higher rates of behavior than girls on Seeks Help from Peer, $F(1,134) = 5.06$, $p < .03$, and Very High Activity Level, $F(1,134) = 4.55$, $p < .04$. Girls scored higher on Negative to Bid from Peer, $F(1,134) = 6.39$, $p < .02$, Talks to Adult, $F(1,134) = 3.12$, $p < .08$, and Wanders, $F(1,134) = 3.41$, $p < .07$. There were no sex differences on the global ratings. On responses to structured situations, boys scored higher than girls on Level of Involvement with Toys in the Novel Environment, $F(1,134) = 7.51$, $p < .01$, and made more Errors of Commission and more Anticipatory Responses, $F(1,110) = 6.21$, $p < .02$, and $F(1,110) = 7.00$, $p < .01$, respectively, on the Impulse Control Game.

On cognitive tests, boys scored higher than girls on Persistence on the Impossible Puzzle, $F(1,134) = 6.67$, $p < .02$. Girls scored higher than boys on Memory for Objects, $F(1,134) = 6.46$, $p < .02$, and Hidden Figures, $F(1,134) = 2.75$, $p < .10$.

4.4.2.3 Supplementation X Sex interactions Of the variables assessed using the Interaction Record, interaction effects were identified for Verbal Aggression, $F(1,110) = 6.00$, $p < .02$, Rough and Tumble Play, $F(1,134) = 3.68$, $p < .06$, Talks to Adult, $F(1,134) = 3.25$, $p < .08$, Distracted, $F(1,110) = 3.30$, $p < .09$, and Very High Activity Level, $F(1,134) = 6.21$, $p < .02$. Examination of the means shown in Table 4–8 indicates (a) supplementation was positively related to Verbal Aggression and to Rough and Tumble Play for boys, but negatively related for girls; (b) supplementation was positively related to Very High Activity Level for boys, while there was no relation for girls; (c) supplementation was unrelated to Distracted for boys but negatively related for girls; and (d) supplementation was positively related to Talks to Adult for boys and negatively related for girls. On global ratings, there were two interaction effects. High supplementation was related to high energy level and positive disposition for boys only, $F(1,134) = 3.32$, $p < .08$, and $F(1,134) = 4.98$, $p < .03$, respectively.

There was one interaction effect with respect to structured situations. On the Impulse Control Game, the supplementation effect for Anticipates Response was specific to boys, $F(1,110) = 4.76$, $p < .04$. On cognitive tests, the one interaction effect was for Digit Span, $F(1,134) = 4.61$, $p < .04$, with supplementation positively related to test performance for boys and somewhat negatively related for girls.

4.4.3 Summary of results

As we had hypothesized, the child's nutritional history was significantly related to a number of important dimensions of social and emotional functioning at school age and, specifically, to interest and involvement in the peer group's activity, activity level, and affective characteristics. Better-supplemented children were more often involved with peers, showed more positive affect, were less often anxious, and were more frequently characterized by a moderate activity level than were less well-supplemented children. In addition, they showed the predicted superiority to poorly supplemented children in responses to moderately stressful situations. They showed more exploration of a novel environment, more persistence on a frustrating task, more involvement in a competitive game, and greater alertness and motor control on an impulse control game. In addition, a clear pattern emerged of supplementation-behavior relationships which were significant only for boys. Supplementation increased physical, aggressive, and high-activity behaviors for boys, while there were negligible or negative effects of supplementation on these behaviors for girls. These differential outcomes are illustrated in Figures 4–11 and 4–12, which show the major supplementation effects (i.e., across sexes) and Supplementation X Sex interactions, respectively, with respect to children's social interaction and emotional characteristics. In addition, sex differences in social-emotional and cognitive behaviors, and Supplementation X Sex interactions with respect to other classes of behavior variables were identified.

One of the interesting implications of the findings, and in particular the Sex X Supplementation interactions, is that they suggest that it may be important to distinguish (a) the general effects of improved energy intake on the child's ability to engage with the social and nonsocial environment, from (b) the specific patterns of behavior that develop. To evaluate better this hypothesis that there are both general and environment-specific effects of nutritional supplementation on later behavior, we consider findings from supplementary analyses on the effects of early calorie supplementation on later behavior within villages.

4.4.4 Supplementary analyses

4.4.4.1 Within-village analyses Within-village analyses were of interest for several reasons: (a) The villages differed in presence/absence of protein supplementation. Were relationships between caloric supplementation and behavior dependent on whether the supplement was protein-rich or protein-free? (b) Both supplemental caloric levels and home-diet caloric levels were significantly higher in the large protein-calorie village than in the calorie-only village over the course of the INCAP Longitudinal Study. Were relationships between calorie supplementation and behavior dependent on absolute level of energy intake? (c) The villages differed on several objective measures of quality of home and social environment. Although the villages originally had been matched on a number of general SES indices and demographic characteristics (e.g., means of livelihood, home diet), analyses based on more recent data show that the villages differed significantly on such variables as value of producer durables, access to water supply, modernity of cooking facilities, and number of rooms in the house. Further, the general atmosphere of the villages differed. For example, San Juan, the smaller, protein-calorie village, was shady and cool, physically atractive, and well cared for. Conocaste, the large protein-calorie village, was, in contrast, extremely hot and depressing, with a harsher, browner landscape, and with little physical beauty. Certainly such qualitative differences might affect the persons living in these environments. Thus, it was important to see whether relationships between nutritional history and later behavior were similar for the different villages.

The within-village analyses shown in Table 4–7 yielded results highly concordant with the across-village results, with early caloric supplementation related to later levels of social responsiveness, affect, and activity.

In each village, however, there were marginally significant relationships between supplementation variables and behavior measures that were not predicted or suggested by the overall analyses. In Conocaste, the protein-calorie village, the addition of CS1 to the equation improved the prediction of Intrudes (i.e., takes materials from peer), with a positive beta weight for CS1, $F(1,42) = 3.84$, $p < .10$. Also, there was a positive though not significant relationship between CS1 and negative response to Bids for Help from Peer. In Santo Domingo, the calorie-only village, there were several unexpected findings. First, child caloric supplementation from 2 to 4 years, rather than the maternal or infancy supplementation measures, was the most consistent predictor of the social-emotional measures. With the other predictor variables in the equation, CS2 was significantly associated with high rates of friendly interaction, $F(1,51) = 7.01$, $p < .05$, Plays with Materials, $F(1,51) = 5.20$, $p < .05$, Imaginative Play, $F(1,51) = 3.40$, $p < .10$, and Happy Affect, $F(1,51) = 2.99$, $p < .10$, and with low rates of No Interaction with Peers, $F(1,51) = 8.96$, $p < .01$, and Low Activity Level, $F(1,51) = 3.26$, $p < .10$. Second, child supplementation from birth to 2 years (CS1) was associated positively with Seeks Physical Contact from Adult, $F(1,52) = 4.87$, $p < .05$, a finding that seems inconsistent with the general finding that early caloric supplementation is related to less attention-seeking from adults. Third, there was a significant supplement effect for Helps Peer: Mother supplementation (MS) was the most important predictor and was weighted positively, $F(1,52) = 5.92$, $p < .05$.

To summarize, there was a tendency for supplement effects to be specific to the more prosocial, affiliative behaviors in Santo Domingo, the village given the calorie-only supplement, and to intrusive, aggressive behaviors in Conocaste, the village given the protein-calorie supplement. Further, in Santo Domingo, supplementation after infancy (i.e., ages 2–4) was a more important predictor of social-emotional functioning at school age than was maternal supplementation during pregnancy or child supplementation during the first two years. This may have been due to psychometric properties of the predictor variables; see Barrett et al., 1982.

4.4.4.1.1 Individual tests With regard to individual test scores, the results of the within-villages analyses were generally consistent with the across-village analyses, indicating that caloric supplementation had significant effects on attentional characteristics. Results are summarized in Table 4–7. In both villages, CS2 was significantly associated with high scores on Persistence on the Impossible Puzzle, $F(1,41) = 4.62, p < .05$, for Conocaste and $F(1,51) = 5.26, p < .05$, for Santo Domingo. In Conocaste, the protein-calorie village, CS2 also was associated with high scores on the Following Directions Test, $F(1,41) = 3.43, p < .10$. In Santo Domingo, the calorie-only village, MS and CS2 were associated with Hidden Figures Performance, $F(1,53) = 8.34, p < .01$, for MS and $F(1,51) = 7.02, p < .05$, for CS2. Results also indicated, however, that in the less well-supplemented village, Santo Domingo, caloric supplementation had significant effects on performance on more strictly cognitive measures. In this village, child caloric supplementation from ages 2–4 (CS2) was related to Performance on Block Design, $F(1,51) = 4.5, p < .05$, Bender-Gestalt, $F(1,51) = 2.92, p < .10$, and Cognitive Composite, $F(1,51) = 5.44, p < .05$.

The finding of significant associations between caloric supplementation and cognitive variables only in the less well-supplemented village (Santo Domingo) is important and suggests that the effects of malnutrition and nutritional supplementation depend on the baseline nutritional status of the population. They are consistent with the position (discussed in the final section of this chapter) that malnutrition is likely to have its most severe behavioral consequences under conditions of greatest social-economic disadvantage, with respect to such characteristics as health of the population, environmental stimulation, and social support.

We may now perhaps better evaluate the interactions concerning supplementation and sex. It appears that the general effect of supplementation on later behavior is to increase the child's ability to seek out stimulation and respond to environmental stimuli, but that the specific behavior patterns that emerge will depend to some extent on the child's social and cultural milieu, including the physical environment, available role models, and behavior norms and expectations. This interpretation is supported both by the significant Supplementation X Village effects and also the Sex X Supplementation effects. The latter support this interpretation, since they show that boys and girls, who are socialized differently and for whom there are markedly different behavior norms and expectations, are affected differentially by early nutritional supplementation. As one might predict, the masculine-type behaviors—physical, aggressive, assertive—appear to be enhanced by supplementation for boys. The data suggest that improved supplementation among girls may increase their ability to control their behaviors, so that 'inappropriate' behaviors (i.e., aggression) are inhibited.

4.4.4.2 Tests of potential confounding variables There were two categories of variables that needed to be examined as potential mediators of the relationships between early caloric supplementation and later behavior, and/or confounding influences. First, maternal background factors reflecting education, intelligence, and life circumstances might be systematically related to supplementation and at the same time predict to child social and emotional development. Such factors are frequently viewed as potential confounding variables in the relationship between malnutrition and behavior (Lloyd-Still, 1976; Rush, 1984). Second, the child's early nutritional history might predict to current nutritional status and, if related to behavior, this variable might be construed as a mediator of supplementation effects.

4.4.4.2.1 Maternal and background factors We examined differences between High and Low Supplementation children on ten maternal variables: mother's age at conception, mother's head circumference, mother's literacy (dichotomous scoring), quality of the house, amount of teaching of preschool children, amount and quality of clothing, and "modernity." Only one variable differentiated the two supplementation groups: Mothers in the High Supplementation group had larger head circumferences than mothers in the Low Supplementation group (means were 51.35 cm for High

Supplementation, and 50.75 for Low Supplementation; $t(132) = 2.40, p < .02$).

We examined the relationship between this variable and each of the behavior variables obtained in the structured situations, separately for each sex. For each sex, of 13 correlations, one was significant at $p < .05$. For boys, maternal head circumference predicted number of strategies on the frustrating task; $r(68) = .29$, $p < .02$. For girls, the association was with level of initiative across three situations: $r(55) = .29, p < .03$.

While the correlations for head circumference suggest a possible role for maternal nutritional history in the prediction of child behavior, the total set of findings involving maternal charcteristics strongly suggests that maternal and other background factors are not as important as child supplementation in predicting child behavior. It should be noted that of the other background variables studied (i.e., not including maternal head circumference), not one showed a significant relationship with a behavior variable that replicated across sexes. In fact, there were few significant correlations overall; of 260 correlations, the number reaching significance was about the number expected by chance.

4.4.4.2.2 Current nutritional status We constructed four measures of weight for height, a frequently used measure of current nutritional status (Waterlow, 1984). The measures were percentile, percent of median, standard score, and standard score derived percentile, all based on the 1976 National Center for Health Statistics normative data (NCHS, 1976). There were no differences between High and Low Supplementation children on any of the weight for height measures. In fact, none of the supplementation variables—mother's supplementation during pregnancy, child supplementation from birth to age 2 years, or child supplementation for 2 to 4 years—predicted weight for height for either sex.

With regard to prediction of behavior, there were no relationships between weight for height and any of the structured situation behaviors for boys. For girls, there was one significant relationship, in the direction opposite to expectation: Heavier girls used fewer strategies on the frustrating task; $r(57) = .30, p < .02$, using the weight for height variable representing standard scores.

These findings support the position that the effects of nutritional

TABLE 4-11

Sample means for height for age, weight for age, weight for height*

	Total sample (n = 137)*	Boys (n = 78)	Girls (n = 59)†
Height for Age			
Percentile	6.78	6.22	7.52
Standard score (z)	−2.22	−2.42	−1.96
Percent of Median	90.26	89.84	90.82
z score-derived percentile	4.60	3.55	6.01
Weight for Age			
Percentile	10.42	9.24	11.99
Standard score (z)	−1.63	−1.80	−1.42
Percent of Median	79.76	78.67	81.20
z score-derived percentile	9.23	7.67	11.30
Weight for Height			
Percentile	45.03	43.25	47.38
Standard score (z)	−.21	−.28	−.12
Percent of Median	98.38	97.77	99.18
z score derived percentile	44.49	42.76	46.77

* Data are based on normative values from National Center for Health Statistics (1976).

† Data were not obtained for one of the girls in the sample.

From Barrett and Radke-Yarrow, 1985.

supplementation on behavior are not "explained" by the effects of current nutritional status.

4.4.4.3 Attained height and weight As noted earlier, previous studies of the INCAP population showed that 85 to 90 percent of preschool age children were at least mildly to moderately undernourished, using the Gomez criterion of below 90 percent of expected weight for age.

Data were collected on height for age, weight for age, and weight for height (see previous section) using the National Center for Health Statistics normative values as reference values. Table 4–11 shows the mean values for percent of NCHS median value for each of the three anthropometric variables. In addition, mean percentiles, standard scores, and standard score-derived percentiles are shown.

The mean percent of median height for age was 90.26 with almost 90 percent of children chronically undernourished, using the criterion of below 95 percent of standard (Waterlow, 1984). On weight for age, the mean percent of median was 79.76 with over 90 percent of the children below 90 percent of median value. In contrast, the mean percent median value for weight for height was 98.38 and just over 10 percent of the children were below the cut-off value of 90 percent of standard (Waterlow, 1984). The results support the view of a chronically malnourished population with significant stunting but not wasting.

As in earlier analyses (Barrett et al., 1982), the child's supplementation during infancy and early childhood predicted linear growth. Correlations between calorie supplementation from birth to age 2 and standard score for height were $r(76) = .27, p < .02$ for boys, and $r(57) = .35, p < .01$ for girls. Correlations using supplementation from age 2 to 4 years were $r(76) = .29, p < .02$ for boys, and $r(57) = .27, p < .05$ for girls.

5.0 MODERATING VARIABLES AND AN INTEGRATIVE FRAMEWORK

In interpreting these findings and considering the implications for the developing child, there are three closely related but distinct interpretive issues:

1. The issue of the processes mediating the nutrition-behavior relationships and, specifically, the social-interactive mediating factors;

2. Passive nutrition-environment covariation—the fact that malnutrition frequently covaries with environmental deprivation; and

3. The nutrition-environment interaction: The differing effects of malnutrition depending on the social-interpersonal environment.

We explicated in the introduction to this chapter a model describing interactive processes mediating the effects of malnutrition on behavior. In severe forms, malnutrition affects brain biochemistry and structure. Even in less severe forms there will be energy depletion and metabolic disturbance. As a result of the nutritional-metabolic effects on the organism, there will be alterations in behavior at the time of the malnutrition. There is evidence that the malnourished infant is poorly equipped to stimulate the caregiver and to respond to social initiations, and thus to develop a normal (i.e., reciprocal and synchronous) pattern of interaction with the caregiver. Malnutrition in the caregiver herself would exacerbate the problem in that she would be unlikely to have the attentional and motivational capacities to be sensitive to the child's characteristics, and the child's needs for care and stimulation.

There is evidence of interactive disturbances in dyads where the child and/or mother is malnourished. Chavez and Martinez (1979), in the study reported earlier, found that more poorly nourished mothers were less responsive to and interacted less with their infants than did more adequately nourished mothers with their infants. Pollitt and Wirtz (1981), in a study of 40 Puerto Rican mother-infant pairs, found that aspects of the mother's feeding behavior in the first two days predicted infant weight gain in the first month. Vietze and colleagues (1980), with a United States population hospitalized for growth failure, and Cravioto and DeLicardie (1972), studying mother-infant interaction in Mexican infants, found significant correlations between maternal attention and responsiveness to the newborn and the child's later growth. The latter studies are particularly important because they indicate not only that the quality of mother-infant interaction differs in well-nourished and malnourished dyads (a finding which has been replicated by several investigators) but, further, that problems in interaction may precede the onset of malnutrition, as well as result from the effects of malnutrition on the infant and/or caregiver.

The second issue is covariation as we generally think of it: 'confounding' or passive nutrition-environment correlation. It often is the case that the degree of malnutrition in a population is correlated with other types of environmental disadvantage, which are themselves associated with later behavioral outcomes (see Christiansen, Vuori, Mora, & Wagner, 1974; Cravioto & DeLicardie, 1972). Included are variables such as low income, inadequate material resources, and physical hardship. This issue of covariation frequently has been treated as a methodological problem which is inherent in malnutrition research (Elias, 1976; Warren, 1973) and one which makes it virtually impossible to theoreticaly link malnutrition with later intellectual development. The argument is that cross-sectional studies of malnutrition generally confound nutritional deprivation with environmental deprivation and experimental studies of nutritional supplementation invariably involve increased social interaction with investigators and/or other subjects.

To the degree that nutrition-environment covariation results in an investigator wrongfully attributing extranutritional effects to the influence of nutritional variables, there would be cause for concern. In most studies, however, this does not seem to have been the case. Most studies which have attempted to statistically separate out the effects of nutritional and extranutritional variables have found important malnutrition effects; for example, the studies of Galler et al. (1983a), Barrett et al. (1982), and Edwards and Grossman (1980). The results reported in the present study (see previous section on potential confounding variables) provide particularly strong evidence of nutrition effects over and above the effects of social-environmental variables. Even when statistical analyses have shown stronger social-environmental effects than nutritional effects—as in the Richardson study—the results have been interpretable rather than ambiguous.

Rather, it seems that much of the criticism directed at malnutrition research is based on a confusion of covariation with statistical interaction. For example, Lloyd-Still (1976), reviewing the research on the effects of malnutrition on cognitive development, noted that only when malnutrition occurred in the context of poverty and environmental deprivation were there likely to be long-term effects on intellectual function. Levine and Wiener (1976a) noted that the behavioral effects which appear when malnutrition in animals is accompanied by decreased handling or stimulation are generally diminished when environmental stimulation is not experimentally reduced. Such patterns of findings have been interpreted as weakening the position that malnutrition per se has long-term adverse effects on behavior.

But what these findings suggest is not that the effects of malnutrition are artifactual—that is, due to extranutritional factors. What they show is that the effects of malnutrition on social, emotional, and cognitive development are *highly dependent* on the caregiving and larger environmental contexts in which they occur. The effects of malnutrition on behavior appear to be mediated by social-interactive factors, reflecting both the organism's changed nutritional status (and, presumably, central nervous system functioning) and the response of the social environment, and perhaps the health status of the caregiver. When any of these variables is modified due to a supportive environment, the potential adverse effects of malnutrition may be prevented, diminished, or even compensated for. Thus, what we have is not a 'confounding' of nutrition and environment, but a true statistical interaction where the probable effects of early nutritional insult are highly dependent on the familial and social-cultural context in which they occur.

The present findings of adverse effects of early malnutrition on child behavior at school age in an economically and socially disadvantaged population are consistent with this interpretation. Further, the finding that effects on cognition emerged, but only for the least well-supplemented village, supports the position that malnutrition may have its most negative effects on behavior when the environment is most depressed.

Recent and dramatic examples of this interaction between nutrition and environment are available in clinical studies of nutritional rehabilitation for previously undernourished children. Major data collections are those by Grantham-McGregor and colleagues in the West Indies, Monckeberg in Chile, and Yaktin and MacLaren in Lebanon.

In a first study, Grantham-McGregor et al. (1978) studied 17 6- to 24-month-old malnourished children and 15 adequately nourished comparison children. On hospital admission, the malnourished children had weights below 60 percent of expected weight for age. Children first were tested within three days after admission. Follow-up tests were one week later, prior to discharge, and one month after discharge. Both groups showed the same increases in developmental quotient (22 to 23 points) over the course of recovery. The malnourished group was 25 points lower to begin with, and was still 24 points lower (mean DQ = 85.8) at the final testing. The authors interpreted this finding as showing that malnutrition results in retarded intellectual development. Gains in the two groups were assumed to reflect increased familiarity with the testing situation.

In a second study, however, some amount of 'catch-up' was achieved. Grantham-McGregor et al. (1980) examined the effect of adding 'psychosocial stimulation' to the treatment of malnourished children. There were three groups of children, again 6 to 24 months of age. One group consisted of 18 children with severe malnutrition who received standard hospital care. A second group was a nonmalnourished control group admitted to the hospital for diseases other than malnutrition. A third group consisted of 21 malnourished children who received both nutritional rehabilitation and a cognitive intervention. The intervention involved a program of structured play with the children one hour a day for six days a week during the period of hospitalization.

Follow-up was at admission, discharge, one month after discharge, and six months after discharge. The malnourished nonintervention and intervention groups had developmental quotients of 61 and 64, respectively, at hospital admission, compared to a mean score of 86 for the comparison group. By six months after discharge, developmental quotients were 82 for the nonintervention group, 96 for the intervention group, and 105 for the comparison group. The findings of early behavior impairment and gain in the malnourished nonintervention group are consistent with the findings from the earlier study. The significantly greater increase in the intervention group illustrates the possibility for catch-up in intellectual development following severe malnutrition in infancy with a combination of nutritional and cognitive stimulation. The continued discrepancy in intellectual attainment between the intervention group and the control group suggests the long-term nature of the behavior impairment, and the relative magnitude of the deficit.

Yaktin and MacLaren (1970) conducted a longitudinal study in Beirut. Subjects were 55 infants suffering from acute and severe malnutrition. Children were between 2-1/2 months and 19 months at the time of hospital admission. During the course of hospitalization, they were tested every two weeks. In order to study the effects of stimulation, half the infants were roomed in a highly stimulating environment, with bright colors, toys, music, and a play area. The other half of the group received routine care. In addition, the stimulated group received additional attention from nurses. A control group of 30 nonmalnourished infants was included in the study.

Both stimulated and nonstimulated malnourished infants improved consistently in developmental quotient over the course of hospitalization. Mean DQs rose from 50 to 65 in the nonstimulated group and 55 to 75 in the stimulated group over the four-month period. In both groups, the greatest improvement occurred over the first three weeks, with mean increases of at least ten points in both groups. The mean gains in the stimulated group exceeded the gains in the nonstimulated group. The differences in improvement were significant by the fourth testing. Both groups, however, had far lower DQs at the final testing than the control group. The control group had a mean DQ of 105 at the initial assessment and 108 at the final testing. Like the results of the McGregor study, the findings show the possibilities for improvement of intellectual function in children who were severely malnourished in infancy through nutritional rehabilitation, but also suggest that there are likely to be long-term intellectual deficits.

In Chile, Monckeberg (1981) has been studying the progress of infants admitted for hospitalization for malnutrition. Infants have been studied under two types of treatment: In one type of treatment, children have been admitted for routine hospital care. In a second type of treatment, a stimulation component has been included. The more comprehensive treatment involves nutritional supplementation, 30 minutes a day of 'psychosensorial stimulation' based on Piagetian concepts, physical exercise for 30 minutes twice a day, and 'affective stimulation' provided by volunteers. Monckeberg reports a study of 80 children getting comprehensive care and 80 children getting routine care. The average age of children at admission was 5 months. Psychomotor developmental quotients on admission were 56 and 55 for the routine care and comprehensive care groups, respectively. Monckeberg found that after four months of treatment there were striking differences between the two groups in psychomotor development. By 150 days of treatment, the pediatric hospital treatment group had a mean DQ of 65, while the more fully treated group had a mean DQ of 85.

On the basis of this study, the program was implemented nationwide. Data on the first 100 children who completed 24 months of follow-up indicated normal values for psychomotor developmental quotient. These results are, in Monckeberg's words, "in disagreement with previous ones that suggested permanent psychomotor retardation as a result of severe malnutrition in early life" (p. 148).

All of the studies on follow-up of severely malnourished children are consistent in indicating serious behavior impairment in the months immediately following hospitalization. They are consistent in showing that under the conditions in which recovery *normally occurs*, the child who has suffered from severe undernutrition in infancy is likely to show continued developmental delays, even with improved nutrition. If the caregiving environment is radically improved, the possibilities for developmental advance are increased.

A 'last word' on the topic is best provided from a recent case study on the recovery of a malnourished child from Grantham-McGregor's sample (1984a). The child was a 9-month-old who, when admitted for hospitalization, was thin, wasted, and stunted, with his weight only 78 percent of expected value. After discharge and for two years, he continued to live with his mother in poor circumstances. At the age of 3 years, and after several attacks of gastroenteritis, he was admitted to a children's home because his mother could no longer afford to keep him. After two and one-half years, he was adopted by a teacher, who also had adopted another boy. There now was a dramatic change in the child's environment. In addition to receiving more personal care, he shared a room with his brother and owned books and toys. Following adoption, there were dramatic changes in weight and in developmental quotient. His developmental quotient at the time of hospital admission was 79 and remained at that level until the time of adoption. Between the ages of 3 and 5 years, his developmental quotient increased to 110, well above the average level. In contrast, the developmental quotients of his malnourished peers were stable at the level of approximately 75 over the entire period. The authors concluded, "These data suggest that the behavior of previously severely malnourished children can be vastly improved with good nutrition and appropriate stimulation. Unfortunately, changes in the environment of the degree reported in this case are not readily available. The challenge of treating children with severe PEM (protein-energy malnutrition) is to find low cost, feasible ways of improving their environment so that they can compensate as far as possible for the insult of severe PEM" (p. 18).

6.0 CONCLUSIONS

The effects of malnutrition on child behavior when the child is raised in an environment in which there is little stimulation and opportunity for interaction with the environment are consistent. The child affectively and socially withdraws from the environment and develops behavior patterns that are not conducive to the development of social skills. In infancy, there is a pattern of lethargy, flat affect, lack of attentional capacity, and poor tolerance for stressful stimuli. By early childhood, the child's relationships with peers are predictably affected, and the child shows difficulty meeting the problem-solving demands of everyday life.

There are data, however, to suggest that such maladaptive behavior may not be the necessary consequence of a history of undernutrition. When efforts are made to provide a changed physical and interpersonal environment, one which prevents or works against the tendency for the malnourished child to 'functionally isolate' him/herself from the social and physical world, long-term behavioral effects may be moderated. Malnutrition as it currently exists in the world is associated with a disturbing and consistent complex of behavior impairments in the infant and young child. Only when in addition to nutritional rehabilitation, there are dramatic improvements in the environmental conditions that accompany endemic undernutrition are we likely to prevent such adverse behavioral outcomes.

Chapter Five: Implications and Conclusions

What are the implications of this research for theory, further research, and public policy? In this final chapter we will address five issues: First, how reliable are the present findings? Are they consistent with previous research, and are they supported by research currently being conducted? Second, what are the processes that are likely to have brought about the effects that were identified? Is the 'interactional' hypothesis the most tenable? Are there other data to support the interpretation that disturbed interaction may be a mediating factor between malnutrition and later behavior impairment? Third, what are the implications of the findings with respect to the critical period concept? Do they strengthen or weaken the position that there is a critical period in human development such that factors interfering with brain development at that time will have irremediable effects on central nervous system structure and function? Fourth, do the results provide documentation concerning the relative vulnerabilities of social versus intellectual competencies in children? Do they suggest greater resiliency or greater vulnerability for certain aspects of behavioral function than others—or, rather, do they suggest that there is an interdependence of social and cognitive function, with both equally vulnerable to nutritional insult? Finally, what are the as-yet unanswered questions that may help to clarify the relationship between nutritional history and child development? Where should future research studies in the area of malnutrition and child behavior direct attention?

1.0 RELIABILITY OF THE PRESENT FINDINGS

In Chapter Three we reviewed a number of studies which suggested that certain dimensions of child behavior would be particularly vulnerable to early malnutrition. In the following discussion we will recapitulate these findings but in a somewhat different context; namely, as studies which replicate, to some degree, the findings of the present research.

Let us begin with the present findings. The results of the research show a significant relationship between undernutrition in infancy and behavior at school age. Chronically undernourished children were characterized by low social responsiveness, poor attention, general lack of affect, low persistence in problem solving, and irregular activity level. In addition, they showed less exploration of and interest in a new environment and less success in a response inhibition task. In sum, social responsivity, attention, emotionality, and activity level were affected by early undernutrition. We noted earlier that research on animals and on human infants had led us to predict such a constellation of behavior problems. We come now to the question: Do other studies of undernourished *children* indicate the same impairments?

There have been five studies that have focused specifically on the behavioral characteristics at school age or later of undernourished children. The findings indicate an extraordinary concordance between studies, all of which focused on different populations and examined children who had suffered varying degrees of malnutrition. A summary of the findings, from Frank (1984), follows. The studies are summarized in Table 5–1.

First, to review the findings from the present study. We examined 138 boys and girls from a population in Guatemala characterized as endemically malnourished. Children were classified with respect to degree of caloric supplementation, for their mothers and for themselves, from conception to age 4 years. They were followed up at ages 6 through 8 years. There were significant relationships between degree of undernutrition, inferred from amount of supplementation, and child behavior. The more undernourished children showed less social responsiveness in group play situations, a less well-modulated activity level, less positive affect, and more anxious

TABLE 5-1

Studies of social-emotional functioning following undernutrition in infancy: sample characteristics and measures employed

First Author	Site	Total N	Gender	Sample Characteristics	Hospitalized for Malnutrition	Age at Follow-up (Years)	Measures Used
BARRETT (1982)	Guatemala	138	78 boys 60 girls	56 well supplemented 29 less supplemented 53 poorly supplemented	No No No	6–8	Structured observations, IQ tests, Height and weight
GALLER (1983)	Barbados	258	154 boys 104 girls	129 children with second or third degree malnutrition in first year of life 129 controls, schoolmates of same age, gender and handedness	Under age 1 No	5–11	Teacher ratings, IQ tests, Height and weight, Arm circumference
STOCH (1982)	South Africa	40	22 boys 18 girls	20 children with «gross under-nourishment» 20 controls, age and sex matched	Under age 2 No	20	Psychosocial interview, IQ tests, CT scan, Height and weight, Head circumference
RICHARDSON (1972)	Jamaica	298	All boys	58 survivors of marasmus or kwashiorkor 31 controls, male siblings of survivors 89 controls, classmates of survivors and siblings	Under age 2 No No	5–11	Teacher ratings
BEARDSLEE (1982)	USA	52	Not recorded	31 children with malnutrition from cystic fibrosis or bowel disease 21 siblings	Under age 6 mos. No	5–22	Teacher ratings, Parent ratings, Psychiatric interview, IQ test

From Frank, D.A. (1984) Malnutrition and child behavior: A view from the bedside. In J. Brozek and B. Schurch (eds.), *Malnutrition and Behavior: Critical Assessment of Key Issues*. Lausanne, Switzerland: Nestle Foundation, pp. 307–326. Reprinted by permission of the publisher.

behavior. In structured situations, they showed less exploration of a novel environment, poorer motor impulse control, and a lack of persistence and initiative in problem-solving situations. They were rated by observers as less attentive, energetic, happy, and self-confident than better-nourished peers.

A second study has been conducted by Galler and colleagues (1983b) in Barbados. Subjects in the study were 129 boys and girls who had been hospitalized for undernutrition before the age of 1 year, with second- or third-degree malnutrition. They were compared with 129 matched controls, schoolmates of the same gender, age, and handedness at ages 5 through 11 years. The malnourished group showed a higher rate of school failure than the control children. They were rated by teachers as more distractible, more apathetic, and more emotional than nonmalnourished peers. They were less socially involved with both peers and adults. They scored more poorly on standardized intelligence tests and evidenced deficits in memory.

A third study was the long-term study by Stoch and colleagues (1982) in South Africa. Subjects in the study, which was first reported in 1963, were 20 children (9 boys and 11 girls) who had suffered from severe malnutrition as infants. All children had been hospitalized before the age of 2 years. They were matched with 20 nonmalnourished children for age, sex, and social background. Follow-up studies began at age 3 years and continued through early adulthood (age 20). Stoch reported that significantly more children from the index (malnourished) group experienced school failure. The previously malnourished children also showed emotional and behavior problems; in particular, a lack of impulse control and social withdrawal.

A fourth study was carried out in Jamaica by Richardson et al. (1972). Subjects were 74 boys all of whom had been severely malnourished before the age of 2 years. They were followed up between the ages of 5 and 11 years. Each index child was compared with a matched-control child; in 38 cases this was a sibling. Subjects were assessed on teacher rating scales for child behavior. Assessments included ratings of behavior problems relating to classwork, social relationships, and conduct problems. The children also were evaluated by peers for popularity. The children who had been hospitalized for malnutrition scored significantly lower than the control children both in behavior relating to classwork and in school functioning. They were rated by teachers as having poorer attention and memory, being more distractible, and as contributing less to classroom activity. They were rated as less cooperative and as less able to get along with peers. Previously malnourished children also were less popular than the control children, as indicated by lower scores on the sociometric measure.

A fifth study differed from the others in that the subjects were hospitalized children in the United States who had suffered from cystic fibrosis or other bowel dysfunction during infancy and thus were acutely malnourished for an extended period of time, but did not experience disadvantageous social-environmental conditions. Beardslee et al. (1982) studied 31 malnourished children, all hospitalized before age 6 months, and 21 sibling controls. Measurements were taken at age 5 years or older, with the oldest subjects followed into early adulthood. There were no differences between the groups in social or emotional characteristics as determined by psychiatric interview and teacher and parent ratings, or in measured intelligence as determined by standardized tests. The malnourished group did show a higher rate of school failure than controls.

The concordance of the results is demonstrated in Table 5–2. With the exception of the Beardslee et al. study, all studies showed that early malnutrition affects the child's attention, emotional control, affect, and, when measured, interaction with peers. Malnourished children show greater apathy and social withdrawal, greater difficulty controlling emotions under normal pressures and stresses, increased distractibility, less persistence, and poorer peer relations than better-nourished children. In addition, there is some evidence of memory impairment, attributed by Galler et al., to attentional deficit, and a higher rate of school failure.

It is interesting that behavioral differences between malnourished probands and nonmalnourished siblings were not found in the Beardslee et al. study, since this was the only study in which children did not experience harsh physical or social-environmental circumstances. In each of the other studies, home environments were characterized by extreme poverty, few educational resources, and a high susceptibility to infection. In addition, in the studies in which significant effects of malnutrition were found, mothers were poorly educated and likely to have had little knowledge of child-

TABLE 5-2

Studies of social-emotional functioning following malnutrition in infancy: outcomes observed

First Author	School Failure	Distractibility/ Inattention/ Impersistence	Memory Impairment	Irritability/ Impulsivity/ Emotional Lability	Apathy/ Emotional Withdrawal/ Sadness	Inactivity	Interaction with Peers	Interaction with Adults
Barrett (1982)	NR	+	NR	+	+	+	−	±
Galler (1983)	+	+	+	+	+	NR	−	−
Stoch (1982)	+	NR	NR	+	+	NR	NR	NR
Richardson (1972)	NR	+	+	NR	+	+	−	−
Beardslee (1982)	+	NR	NR	NR	NR	NR	0	NR

\+ = increased in those malnourished in infancy
− = decreased in those malnourished in infancy
± = varied from village to village
0 = no difference
NR = = not recorded

From Frank, 1984.

ren's developmental needs. However, Beardslee and colleagues' work cannot be considered incontrovertible evidence that infantile malnutrition in a socially privileged environment has no adverse socioemotional sequelae, since the control group, siblings of chronically ill children with a fatal genetic disorder, are themselves prone to psychosocial disorders (Lavigne & Ryan, 1979). A high level of psychiatric morbidity in the controls (3/21 had major emotional difficulties) may obscure the relationship between nutritional history and later social-emotional functioning.

2.0 PROCESSES MEDIATING THE RELATIONSHIPS BETWEEN MALNUTRITION AND BEHAVIOR

A second question concerns the processes mediating nutrition-behavior relationships. How does malnutrition affect later behavior? We referred earlier, in Chapter Four, to an interactive model describing how the malnourished infant is less successful at eliciting maternal response than the nonmalnourished infant and, thus, receives less optimal social and physical stimulation from the caregiving environment. There are, as has been shown previously, data to show that the malnourished child does behave differently and interact differently with caregivers than the nonmalnourished child. In addition, we have seen that early malnutrition is predictive of behavioral problems at school age and later.

The question is: Are the early interactive disturbances the processes which account for later behavior problems in the malnourished children? The longitudinal data of Stoch and Smythe and the Jamaica researchers show the persistence of adverse effects of malnutrition on psychological functioning. But data are not yet available which link interactive failures which are themselves due to malnutrition to later behavioral problems.

Given the lack of research to provide direct support for this interactive model, credibility hinges largely on evidence that interactive processes in infancy *do* relate to later social competence; that the dimensions of mother-infant interaction which have been shown to be influenced by infant undernutrition are indeed related to the kinds of social behaviors and competencies which we have found, in the present research, to be related to early malnutrition.

On this question there are data, and perhaps the most relevant data come from research on the long-term consequences of prematurity. Premature birth may inflict more severe biologic deficits upon the child than undernutrition following term birth because prematurity generally involves a number of serious medical problems such as respiratory distress syndrome, metabolic disorders, and gross violations of central nervous system functioning. Nonetheless, studies that examine the nature of early interaction for such high-risk infants are relevant to the present research issue; particularly, if they relate individual differences in organic impairment to early interaction, which, in turn, may be related to later behavioral outcomes.

One long-term study is of particular interest. Sigman et al. (1981) examined the role of postnatal medical complications in predicting developmental progress at two years. They also examined the relative adequacy of different models that posit alternative pathways for how postnatal complications may affect 2-year functioning. Specifically, they examined the mediating roles of (a)pediatric complications in infancy, and (b)interactive disturbances in early infancy.

Subjects were 126 preterm infants with a gestational age at birth of 37 weeks or less and birthweight under 1500 grams. Medical complications were assessed using a number of separate scales; one for obstetric complications, another for illness and medical events in the first two years. Pediatric complications were assessed for months 2 and 4 through 9 and mother-child interaction was examined at 4 and 8 months. The latter was scored using a measure for interactive risk, based on maternal responsiveness/unresponsiveness and other behavior dimensions. The child's developmental status was examined at 9 months, 24 months, and 5 years, using the Gesell developmental scales and child intelligence tests. The data have been analyzed up to 2 years of age. The data show that postnatal complications alone did not predict to the developmental outcomes at 24 months. Postnatal medical complications, however, did predict to both pediatric complications at 1–4 months and interactive problems at 4 months. These, in turn, predicted pediatric disorders at 4–9 months and interactive disorders at 8 months. Finally, both 4–9 month pediatric complications and the 8-month interactive disorders predicted to 2-year developmental retardation. Thus, the hypothesized relations between early organic impairment, interactive disturbances, and later developmental outcomes were obtained. It is interesting to note that the strongest paths (i.e., causal influences) from postnatal complications to developmental impairment at 2 years were those mediated by interactive disturbances.

A recent study of 1-year-olds also implicates individual differences in maternal behavior in infancy to the child's later behavioral development. Bradley and Caldwell (1984) administered the HOME test of the home environment to 49 lower-SES families when children were 12 months old, and followed-up 37 children at first grade on measures of school achievement. There were strong associations between maternal involvement with the child and acceptance of the child, assessed from direct observation of mother-child interaction at 12 months, and the child's school achievement at first grade.

We currently are conducting a study at the Brigham and Women's and Beth Israel Hospitals in Boston which we hope will further elucidate this issue.* The study examines the long-term sequelae of intrauterine growth retardation (IUGR), measured by birthweight and ponderal index (PI). In this study, 70 children identifed at birth as IUGR are compared with 70 appropriate for gestational age children. Half of the children in each group are assigned to an intervention which occurs during the first month of life and is designed to sensitize the mother to the child's developmental—and idiosyncratic—characteristics. It is predicted that the intervention will prevent the otherwise adverse consequences of IUGR on both mother-infant interaction and child development. As children and mothers are examined beginning at 4 months and extending through age 2 years, and as behavioral measures will focus on precisely those aspects of child behavior (exploration, attention, persistence, affect) known to be affected by undernutrition, the study will provide important information on the short- and long-term consequences of prenatal undernutrition and the mechanisms by which early organic vulnerability may affect later behavior and development, with and without environmental intervention.

* Principal investigator for this study is T. Berry Brazelton. The first author is a co-investigator.

3.0 CRITICAL AND SENSITIVE PERIODS

Perhaps no other concept in this area of study has received as much attention, or has so often been a subject of disagreement, as has the concept of 'critical period'. A major question is whether there is a critical period such that undernutrition occurring during this period has lasting and irreversible effects on development. The corollary is that undernutrition of later onset will have less significant and lasting effects. Before addressing this question, however, it is important to recognize that there is disagreement and misunderstanding concerning the meaning of the term 'critical period'. The term 'critical period' refers, in the broadest sense, to a period of development such that if certain events do not occur at this time, certain later developments are inhibited or disturbed. The term is credited to the field of ethology and has been popularized in the works of Lorenz (1957). Lorenz demonstrated critical periods in a number of animal species for imprinting, the process whereby an infant becomes attached to another animal or thing and treats it with clinging and/or following. The neuroanatomic implications of the critical period hypothesis are outlined in detail in Chapter Three.

There is misunderstanding with regard to the terms critical period and sensitive period. The distinction is important. Critical period refers to a time where if certain experiences do not occur, certain later developments cannot occur. For example, if there is a critical period for language experience, this would mean that if one does not have language experience at that time, normal language development will not occur. In contrast, the concept 'sensitive period' does not imply a necessary phase for certain experiences but, rather, an optimal one. Again, language learning provides an example. Let us consider second-language learning. A child who is exposed to a foreign language in addition to his/her own language from, say, birth to age 3 years, will become bilingual and experience no difficulty learning the syntax of both languages. In later childhood, however, after, for example, the age of 5 or 6 years, the child will have greater difficulty acquiring the second language (given the same opportunity to do so) and by age 10 or 11 years will experience the acquisition difficulties of most adults. Thus, one might posit a sensitive period for dual language learning. It is not that a second language cannot be mastered if experience is provided after this sensitive period; it will, however, be more difficult.

Different perspectives regarding the effects of malnutrition on later behavioral development spawn different models concerning critical and/or sensitive periods. In the 1950s and 1960s a number of researchers, primarily those working with animals, viewed behavior impairment as resulting primarily from organic deficits. According to this view, the crucial question was whether there was nutritional insult during a period when the brain cells still were increasing in number. If nutrient intake was reduced drastically during a period of rapid cell increase or hyperplasia (Coursin, 1972, 1975), it was assumed that central nervous system development would be permanently affected and, thus, that behaviors reflecting central nervous system integrity (such as information processing and intellectual functioning) would be adversely affected. Dobbing (1984) gives a reassessment of this view.

Even among those who have put forth the view that there is a critical period—with respect to malnutrition—for brain growth and central nervous system functioning, there is disagreement about just when this critical period might occur. In any species there is a period of neural growth when the majority of brain cells are replicating (hyperplasia). In humans, this period extends from prior to birth through the end of the first year. It has been suggested frequently, therefore (cf. Herzig et al., 1972), that the critical period for nutrient intake would be prior to birth (beginning in the second trimester) and through the first year of life, particularly through the first nine months.

There is recent evidence that much important central nervous system growth occurs in the second and third year as well, after neural cell number is largely established. Specifically, there is increased complexity of myelinization, synapse formation, and dendritic arborization (Epstein, 1978). We know also that brain growth continues in this way into the first four years; in fact, it has been suggested (Lenneberg, 1969) that the second through fourth years in particular are central for language experience and development. Thus, not only is there a question of whether the critical period for brain growth (and environmental input) is 'critical'or merely 'sensitive'; there is the more basic problem of defining the time period which should be deemed critical or sensitive (as the case may be). A strong case can be made for considering any of the early periods of human development as critical, if the sole criterion is whether or not significant change in neural complexity or structure is occurring. The fact that the functional correlates of such changes are as yet ill-defined makes the matter even more problematic.

A second model, which might be called a transactional model, also has been evolved largely by animal researchers. This model suggests that the development of brain structure and function may be influenced not only by the availability of nutritional substrates but by environmental experience. Central to this view is the recognition that nutritional insult may affect the organism's behavior by interfering with brain growth and function. It also is recognized, however, that some of the effects of malnutrition on behavior are mediated by environmental stimulation or deprivation. Much of the formulative research has been conducted in the area of malnutrition and behavior (Levitsky & Barnes, 1972), but important contributions also have come from the research of Harlow and colleagues on the effects of early isolation on later behavior in primates (Harlow, 1965; Harlow et al., 1968) and research on the effects of environmental stimulation and deprivation on neural development in rats (cf. Rosenzweig & Bennett, 1972).

As noted earlier, Levitsky and Barnes have shown (1972) that severely malnourished animals, including rats, pigs, and dogs, tend to 'functionally isolate' themselves from environmental stimulation, showing fear of novelty, reduced social responsiveness, and reduced exploration. This is a result of their inability to process and act upon sensory input which would be manageable for a more adequately nourished animal. This functional isolation leads to an even lower level of environmental stimulation, and the animal becomes even less competent at responding to physical stress, social stimulation, and other types of environmental challenge. Levitsky and Barnes also found, however, that the deleterious effects of malnutrition on behavior could be reversed or at least compensated for by providing the animal with rich social experience. They showed, for example, that when the malnourished animal was raised with a socially experienced and nonmalnourished 'aunt', the animal tended not to show the nonpurposive and dysfunctional behavior often associated with malnutrition. In fact, the researchers found that the adverse effects of malnutrition *depended* on a corresponding lack of social stimulation. When stimulation was not reduced, as it generally is in the experimental malnutrition research (see Levine & Wiener, 1976a), the adverse behavioral effects were either less extreme or not apparent.

Rosenzweig and colleagues were the first to report on the specific changes in brain structure that could be brought about by increased handling of rats. They showed that diminished handling prevented normal brain cell replication when the environmental deprivation occurred during the period of hyperplasia and that increased handling increased cell number. Effects of environmental deprivation appear to parallel those of decreased nutrient intake in the development of central nervous system structure.

Harlow's contribution was to show that social isolation adversely influenced both behavior toward the mother (i.e., such that the infant never developed a normal attachment relationship to the mother) and behavior with peers. For example, socially isolated monkeys failed to develop normal sexual behaviors. Further, if an early deprived female was forcibly impregnated, the animal was likely to show, as a parent, disturbed mothering behaviors, including abusive and neglectful behaviors toward the child.

In sum, the transactional approach views the organism as vulnerable in early life, that is, during the period of brain cell replication, to impairment of central nervous system structure and function as a result of both the effects of malnutrition and the effects of environmental deprivation. When the two occur at once the infant is doubly at risk. But even if malnutrition occurs in an environment which is not inherently deficient, there may be environmental deprivation nonetheless, either because the infant isolates him/herself from the environment or because—due to his or her behavioral characteristics—the child is less often sought out for social interaction.

The implications of this synergistic model can be made explicit:

1. There *may* be a critical period in brain development when a lack of nutrients to the brain, combined with a lack of environ-

mental stimulation, will have permanent effects on the organism due to deficits and distortions in the attained structure of the brain.

2. Whether or not there is a critical period such that failure to obtain adequate nutrients and/or stimulation permanently affects later functioning, *individual differences* in behavioral capacities still may be discerned, and these will depend on both the degree of nutritional deprivation and the degree to which the nutritional deprivation is correlated with or brings about lack of appropriate environmental stimulation.

3. Even beyond the period of rapid increase in brain cell number, stresses to the central nervous system will have an effect on later behaviors of the organism to the extent that (a) they predispose the child toward certain temperamental or behavioral characteristics (e.g., fragility, hyperalertness, distractibility) and (b) such effects influence the quality of the caregiving.

4. To the extent that behavior impairments resulting from insufficient food energy and environmental deprivation represent behavioral *adaptations* (i.e., ways of responding to the environment which least compromise the integrity of the organism) other, more adaptive ways of responding to the environment can be fostered (i.e., learned). The limits of such learning will depend on the physical capacities of the organism.

What would constitute tests of a critical period model? From a research perspective the original critical period hypothesis predicts that children who suffered from severe malnutrition at a particular period in development would shown impaired performance on any tests of cognitive functioning which depend on central nervous system integrity and on measures of social behavior which depend on higher-level cognitive processes. However, this model is too narrow to account for the data. For example if a child were identified who showed a nutritional history which would be expected to be debilitating but nonetheless attained normal intellectual function—and such events have been documented—the concept of a critical period for malnutrition would have to be modified. That is, one would have to infer that otherwise adverse effects could be inhibited by other factors. One then must posit a moderating influence, which brings us to the second model.

The second model would predict that children who experience

malnutrition in early infancy would show long-term deficits in cognitive processes or in social-emotional functions, but individual differences in severity of deficit would be expected, with better development in children (at a given level of nutritional deprivation) with more supportive social environments, and, in some cases, complete 'compensation'. As with the first model, however, different critical periods would be possible—i.e., prenatal, early infancy, later infancy, or early childhood. Further, specific models would differ in the assumptions made about levels of undernutrition and/or environmental deprivation necessary to bring about significant behavior impairment.

3.1 Systematic Tests

Several studies have addressed the issue of critical or sensitive periods. Cravioto and Robles in Mexico (1965) examined the relationship between timing of hospitalization for kwashiorkor and later IQ. Hertzig et al. (1972) in Jamaica studied the association between time of hospital admission for severe malnutrition and later developmental quotient. Chase et al. (1970) studied the effects of malnutrition on United States children from a deprived population and examined the relationship between age at hospitalization for malnutrition and later intelligence. These three studies may be considered as a group, since they each focused on acute, clinical malnutrition.

Other studies were concerned with chronic undernutrition. In the data analysis by Klein et al. (1977), the importance of prenatal versus postnatal nutritional supplementation on later cognitive development in Guatemalan children was examined. Using a subset of that sample, we (Barrett et al., 1982) examined the contributions of prenatal supplementation, supplementation in infancy, and supplementation in early childhood to later cognitive, social, and emotional functioning.

A third study was our research in San Diego (Barrett et al., 1981) which was conducted simultaneously with the Guatemala study. In this study, in which we used maternal reports of mother and child undernutrition, we examined the relative importance of undernutrition during pregnancy, undernutrition during infancy, undernutrition from age 2 to 6 years, and current nutritional status in predicting cognitive, social, and affective characteristics at first grade.

Fourth, there is the Bogota study, which examined the importance of nutritional supplementation at different periods and for different lengths of time during infancy and early childhood for later cognitive and social development (Herrera et al., 1980).

Finally, Hicks et al. (1982) recently examined the differential effects of WIC nutritional supplementation beginning prenatally or beginning after age 1 year on the cognitive and behavioral development of Louisiana schoolchildren.

3.1.1 Acute malnutrition

Cravioto and Robles studied 20 children who had been hospitalized for kwashiorkor over a period of approximately six months. Six of the children were less than 6 months of age at the time of hospital admission; nine were between 15 and 29 months; five were between 37 and 42 months. The Gesell tests for motor, adaptive, language, and personal-social development were administered to each subject every two weeks, from the time of admission to the end of hospitalization. To determine the relationship between recovery of adaptive and motor skills and time of onset of illness, Gesell test scores were plotted as a function of days of hospitalization for each of the three groups of children. Results indicated that for each of the four categories of Gesell scores, there were marked differences in the slopes of the recovery curves among the three groups. Recovery curves were steepest (i.e., most optimal) for children who had entered the hospital between ages 37 and 42 months, and flattest for those who had entered prior to 6 months. Further, in the two groups admitted to the hospital after 15 months, differences between the performance levels of the children in the sample and the expected performance levels for same-age normal children diminished over the course of hospitalization. The authors interpreted their findings as indicating that protein-calorie deprivation suffered before the age of 6 months is more likely to result in permanent loss of intellectual ability than is deprivation occurring at older age levels.

Hertzig et al. (1972) in a study cited previously compared the intellectual performances of Jamaican children ages 5 to 11 with histories of severe malnutrition with those of sibling controls and nonsibling matched controls. Like Cravioto and Robles, the investigators used age of hospitalization to estimate timing of the nutritional insult. The age range of the children at the time of hospital admission was 3 to 24 months. The authors found that correlations between age of admission and WISC IQ were nonsignificant for full scale and performance and verbal IQ. Moreover, when children who entered the hospital before age 8 months were compared to those entering after 8 months, no significant differences in IQ were found. Thus, the Hertzig et al. data did not support Cravioto and Robles. Timing of malnutrition did not appear to be correlated with later IQ.

The third study, by Chase and Martin, gave still different findings. In their study, cited previously, age of first hospitalization was negatively correlated with developmental quotient at age 3–1/2 years. The children who had entered the hospital prior to 4 months of age all had developmental quotients over 80, in contrast to those admitted after 4 months of age, whose mean developmental quotient was 70.

In sum, the three studies provided entirely inconsistent findings. Cravioto and Robles found that Mexican children admitted early for hospitalization showed the greatest cognitive deficits. Chase and Martin found that children admitted early had the least deficits. And Hertzig found no relationship between time of admission for hospitalization (and, presumably, time of malnutrition) and severity of the cognitive deficit.

Methodological features of each of the studies raise interpretive issues. First, in the Cravioto and Robles study, sample sizes were extremely small, precluding a meaningful statistical analysis of differences in the recovery curves for the three age groups. Second, there were no overlapping data points across the three groups. At the time of the last measurement, the oldest child in the youngest group was approximately 12 months of age, which is *younger* than the youngest child in the middle age group at the beginning of the study. Similarly, at the time of the last measurement, the oldest child in the middle group was younger than the youngest child in the oldest group at the beginning of the study. Since there was no common age level at which the three curves can be compared, one cannot eliminate the possibility that group differences were due to time of measurement and/or that cohorts represented different populations to begin with. Finally, as Hertzig et al. have noted (1972), there are no follow-up data beyond the brief six-month period.

The Hertzig et al. study also raises interpretive issues, all noted by the investigators themselves. First, the failure to establish a relationship between time of onset of illness and later intellectual performance may have been due to factors unrelated to nutritional status; for example, differences between older and younger children in the vulnerabilities to the presumably detrimental effects of hospitalization, or differences in the severity of illness. Second, it may be that the critical period notion was originally too narrowly conceptualized by the investigators. In interpreting their findings the authors suggest the importance of later infancy, beyond the period of neuronal cell division, as important for later brain development. They note that there are structural features of the nervous system, for example, dendritic arborization and synaptic formation, which after the early months of life continue to play a significant role in the evolution of the functional integrity of the nervous system and later psychological functioning.

The interpretations of the Jamaica findings by Hertzig et al. raise an important question about the effects of institutionalization on later mental development. Chase et al. also consider the influence of hospitalization in interpreting their results, but from a different perspective. Chase et al. interpret their findings of a negative relationship between age at admission for hospitalization for malnutrition and later mental development by assuming that late admission indicates prolonged illness without treatment. Thus, those entering latest were those who suffered longest from malnutrition. Interpreted in this way, the negative correlation between age at admission to the hospital and level of mental test performance in early childhood is not inconsistent with other findings (since no data were provided in the other studies on the correlation between severity of initial illness and later mental development); neither, however, do the data speak to the issue of critical periods.

Three other studies of the relationship between time of hospital admission for severe malnutrition and later intellectual development give mixed results. Cabak and Najdanvik (1965) studied the relationship between time of hospitalization for manutrition and intellectual development at ages 7 to 14 for lower-class Yugoslavian children. There was no relationship. McLaren et al. (1973), in their study of Lebanese children, found no association between time of onset of malnutrition and IQ at 3 to 5 years of age. Finally, Evans et al. (1971) in a study in which overall differences between previously malnourished children from South Africa and their siblings were not statistically significant (although both groups were below average) did find that within the malnourished group, children hospitalized for malnutrition between the ages of 10 and 15 months performed more poorly at 9 to 14 years of age than did children hospitalized between the ages of 16 and 48 months. The authors interpreted the findings as reflecting the poorer social environments of the early-onset malnourished group.

3.1.2 Chronic malnutrition

Three studies on chronic malnutrition have used the same general approach to examine the importance of nutrition at different periods in development. The studies are those by Klein et al. (1977), Barrett et al. (1982), and Barrett et al. (1981). The method in each was to examine a population about whom there was information concerning nutritional history at different phases of the mother's and child's development. In each study, the relationships between nutritional variables at the different periods of development and measures of later psychological functioning were determined, and the relative magnitudes of the different relationships examined.

The results of our Guatemala research (Barrett et al., 1982) were reported in detail in Chapter 4, and we will not review all of them. We will mention only the general finding that across villages and sex groups, for an endemically malnourished population, nutritional supplementation in infancy (i.e., supplementation to the child) tended to have a stronger relationship with later behavioral outcomes than did supplementation to the mother before the child's birth. In addition, both the prenatal supplementation and early infancy supplementation variables tended to be more strongly related to the behavioral outcomes than did a variable representing supplementation after the age of 2 years.

Klein et al., examining the relationship between nutritional supplementation and cognitive development on the larger sample of which our study group was a subsample, obtained different findings. Klein and colleagues examined the relationship between total maternal supplementation and cognitive development at preschool age, controlling for postnatal supplementation from birth to 18 months, and the relationship between postnatal supplementation and cognitive development, controlling for prenatal supplementation. In contrast to our findings, Klein et al. found stronger prediction from the prenatal variable. There are at least four possible explanations for the differences in findings. One, of course, is sampling error; Klein et al.'s larger sample enhances the credibility of their findings. There were two other obvious methodological differences between the studies. The age group we examined was 6 to 8 years, as opposed to preschool age. It is conceivable that the nature of the effects (i.e., the causal pathways) changes over time, with some variables increasing in importance as predictors and others decreasing. Perhaps, most importantly, our findings concerned social and emotional characteristics, while Klein et al.'s concerned cognitive outcomes. In those analyses in which we did examine the prediction of cognitive functioning, neither prenatal supplementation nor supplementation in infancy were strong predictors; rather, social-environmental conditions in the home were predictive. There was a fourth and more subtle difference between the two analyses and that was in the way that postnatal supplementation was defined. In the Klein et al. study, the variable representing postnatal supplementation was constructed from measures of both the child's supplemental caloric intake during this postnatal period and the mother's supplemental caloric intake during lactation. In our study the postnatal variable was based solely on child caloric supplement intake. This difference, no doubt, resulted in differences in the correlations between the two predictors (prenatal and postnatal caloric supplementation) and the correlations between the predictors and the outcome variables. It would also help to explain the differences in the results of the two multiple regression analyses.

The third study was carried out in San Diego just prior to the

Guatemala study. Subjects were 65 children, ages 6 to 7, who were participants in the Child Health and Disability Prevention Program (CHDP), a citywide program for lower-income children which provided free health services in the public schools. Children were observed in six-person groups to examine social interaction characteristics and were tested individually for intelligence and school achievement. Mothers were questioned about their own health histories and about the child's health history and current health. The emphasis was on the adequacy of their own and the child's food intake.

There were two major findings. First, with regard to cognitive variables, there was no evidence that the mother's early nutritional history or the child's nutrition during infancy contributed to individual differences in intelligence test or achievement test performance. There were strong correlations, however, between the child's test performance in first grade and the child's recent health history; specifically, illnesses or eating problems interfering with nutrient intake during the years 2 to 6. It is likely that health problems during this period interfered with school attendance and resulted in motivational problems in the child. Thus the 'effect' of food intake problems during this period might be construed as an effect on motivation and learning rather than on intellectual ability. That is, the poor health might have resulted in reduced learning time and motivation to learn, rather than directly impairing intellectual capacity.

Second, there were significant findings regarding the effects of restricted maternal food intake on social and emotional functions and these mirror the findings from our Guatemala study, with respect to the functions affected. Children whose mothers experienced restricted food intake during pregnancy showed less positive affect and less friendly interaction, and were less involved with other children than were childen who had better prenatal nutritional histories. These variables were not predicted by social-environmental conditions. The finding that the prenatal variable was the most effective predictor of the social and emotional outcomes is at variance, to some degree, with the findings from Guatemala. The difference in the relative importance of prenatal versus early postnatal nutrition in the two populations may be related to the differences in study designs and population characteristics. Children in the Guatemala study were from a chronically deprived population and almost all children were nutritionally at risk. In such a population, individual differences in calorie intake during any of several periods—prenatal, infancy, toddler—are likely to have an impact on the growing child and affect physical development and behavior. In the San Diego study, most children were not undernourished. The median heights and weights in the population (generally between the 25th and 50th percentiles across age and sex groups) would not identify the population as small or hungry. In this kind of population, it may be the case that more discrete events, interfering in substantial ways with food intake and thus with the growth and behavior of the child, play a more salient role developmentally, with individual differences in daily food intake over time of lesser importance (i.e., since the population as a whole is receiving generally adequate nutrition). The data relating maternal food restriction to child social development are consistent with this interpretation. The effects, however, are modest and replication would be helpful.

A fourth study which provides data on timing of undernutrition is the Bogota, Colombia study. Waber et al. (1981) compared four groups of subjects that had received varying amounts of protein-calorie supplementation, as well as two additional nonsupplemented groups. One group of supplemented subjects received a family food supplement (623 kilocalories and 20 grams of protein per week) during the first six months of the child's life; a second group from 6 months to 3 years, continuously; and two other groups throughout the first three years of life. (These groups differed with respect to presence or absence of an early education intervention component.) The groups that were supplemented beginning at birth also received prenatal maternal supplementation while the group beginning child supplementation at 6 months did not receive prenatal supplement.

The results for analyses of the effects of supplementation on child development as assesed by the Griffiths test of mental abilities indicated that effects of supplementation during pregnancy and during the first six months of life were short lived. When supplementation was stopped at 6 months, the performance of the early-supplemented group was no better than the performance of the unsupplemented children. Children supplemented throughout the first three years performed better on the tests than children in the nonsupplemented or early supplemented groups. Furthermore, children supplemented beginning only after 6 months performed as well as the children supplemented from before birth to age 3 years. The study did not support the hypothesis that the first six months of life are critical with respect to the effects of food supplementation on later behavioral development.

Although not directly addressing the critical period issue, the results of the Cali, Colombia study (McKay et al., 1978) are relevant here. Children aged 3 to 7 years were included in a preschool intervention program which was both a nutritional intervention and an educational intervention. Children were selected on the basis of height and weight, clinical signs of malnutrition, and per capita family income. There were seven different groups, six low SES, and of these six, five received the experimental intervention. The nutritional supplementation program was designed to provide 75 percent of recommended daily calorie and protein allowances. Supplements were given daily ad libitum. The educational treatment was a program administered to small groups of children and designed to develop cognitive, language, social, and psychomotor skills. Daily health care also was provided. Intervention groups differed in timing and duration of treatment.

Treated children showed significant improvements in cognitive abilities. The most dramatic increases after one year of treatment occurred for children who began intervention at age 3 years, with lesser initial gains at age 4 and still lesser gains at age 5. In all cases, nine months of intervention beginning prior to primary school produced significant increases in performance on standardized cognitive tests. The authors concluded that "environmental deprivation of a degree severe enough to reproduce chronic undernutrition manifested primarily by stunting strongly retards general cognitive development, and that the retardation is less amenable to modification with increasing age" (McKay et al., 1978, p. 277). Treatment effects persisted through 8 years of age, the time of the final measurement.

A final study examined the effects of the WIC supplementation program on the cognitive development of United States schoolchildren (Hicks et al., 1982). There were two groups of children. One group began participation in the program prenatally, during the mother's third trimester of pregnancy, and continued for an average of four years. The second group began supplementation after age 1 and continued for an average of three years.

The subjects in this study were sibling pairs from rural Louisiana. For each of 27 pairs, one child was included in the early supplemented group and one in the later supplemented group. All subjects met nutritional risk criteria with respect to frequency of pregnancy, age at conception, or prepregnancy weight for mothers, and with respect to anemia or weight for height for children. All subjects were lower SES.

In the WIC program, food supplements for pregnant women were milk, eggs, cheese, iron-fortified cereal, and vitamin C-rich juices. For infants under 1 year, iron-fortified formula, iron-fortified cereal, and vitamin-C rich fruit juice were provided. For the older children, the dietary supplement was the same as for the pregnant women.

The authors found that at school age the perinatally supplemented group had higher IQ scores and fewer behavior problems as well as better school performance than the late-supplemented group. The authors attributed the differences in later cognitive performance to differences in the timing of the nutritional supplementation, stating, "...it is probably important that the timing of the nutritional supplementation varied between the two groups, corresponding with the brain growth spurt for the Early Supplement group" (p. 1115). They noted also that the groups differed in total supplementation, with the early supplemented group receiving more supplemental food and over a longer period of time than the later-supplemented group. In addition, they pointed out that the WIC program does not consist only of nutritional rehabilitation; medical monitoring and nutritional counseling are also provided.

3.1.3 Integration of findings

The picture that emerges from these studies is not a clear one. Of the six studies on acute malnutrition in the early years of life, three showed no relationship between apparent age of onset and later impairment; two showed the predicted negative relationship between age of onset and later impairment; and one showed a positive relationship between age of onset and later impairment. Of the studies on chronic undernutrition, three suggested the importance of early maternal malnutrition or supplementation for later behavior; one showed the importance of child nutrition in the first two years; one indicated no relationship between timing of supplementation and later outcomes; and one yielded conflicting findings, with different nutrition variables predicting different outcomes.

It seems desirable to attempt to integrate these findings. First, however, it is important to recognize the limitations of the studies in terms of design and analyses. There are three very obvious considerations. First, studies on the effects of supplementation address a different question from those that compare children who have suffered from discrete episodes of undernutrition. As indicated earlier, the supplementation design addresses a question about the possible effects of nutritional enrichment and presumably allows inferences about the meaning of a corresponding lack of nutrients of the same type at the same developmental period. In fact, however, such an inference requires a certain leap of faith. To take an example, one would not assume that the effects on performance of an athlete's weight gain of 10 pounds would be isomorphic to the effects of 10 pounds of weight loss on performance from the higher weight; the effects (positive or negative) would depend on the period of adaptation to the lighter or heavier weight, *and* the manner in which the weight gain or loss occurred. Similarly, we cannot assume—although logically it seems at least a plausible assumption—that the positive difference in performance obtained by subtracting mean scores on an outcome variable for supplemented minus nonsupplemented children would equal the negative effects on performance which would be obtained if we compared poorly nourished to well-nourished children; i.e., with nutritional status indexed by individual standing on the nutrient being used as the supplement. The effects of adding a positive nutrient are not necessarily equal to the effects of subtracting that nutrient.

Second, in the Guatemala supplementation studies, supplementation began prior to birth for all children. Thus, even though in this analysis one can say whether supplementation at a later time is more important than supplementation at an earlier time (or the converse) with regard to later development, one can say this only with regard to a population supplemented from before birth. One cannot say how important postnatal supplementation would be if begun postnatally (e.g., as it was in the Colombia studies).

Finally, none of the supplementation studies allows us to identify precisely the degree of undernutrition necessary to produce a particular amount of behavior impairment. In each of these studies, undernutrition was inferred from relative standing on a continuous variable and is not a clinical judgment; similarly, behavior impairment is an inference drawn from relative standing on continuous outcome variables.

Given the many methodological differences between studies and the methodological limitations of the studies, we still may draw some broad conclusions. First, individual children are differentially vulnerable to nutritional stress in early life, and different environments—intra- and interfamilial—provide different degrees of protection.

Second, all other things being equal, early malnutrition places the child at risk for mental impairment, that is, significantly greater risk than if no undernutrition occurred. Finally, it is not possible to determine, on the basis of current research, whether prenatal maternal malnutrition or postnatal malnutrition (i.e., alone) during infancy is a greater threat to the child's well-being. Generally, child malnutrition has been evaluated in ways which either confound it with prenatal maternal malnutrition (e.g., as in supplementation studies where maternal caloric intake is correlated with postnatal child caloric intake) or which do not allow us to examine the effects of child malnutrition for children of unequivocally well-nourished mothers. The recent studies do not clarify this issue.

4.0 RELATIVE VULNERABILITIES OF SOCIAL AND EMOTIONAL VS. COGNITIVE FUNCTIONS TO UNDERNUTRITION

While there is much evidence that severe acute undernutrition places the child at risk for mental retardation (Galler, 1984), the effects of chronic undernutrition are less clear cut. The studies cited in Chapter 3 and summarized by Frank (1984) indicate the consistency of relationships between undernutrition and social and emotional functioning. The most recent report, however, summarizing data on the effects of chronic undernutrition on cognitive development showed a much more equivocal pattern of outcomes (Grantham-MacGregor, 1984b). Let us consider the findings from the present study, reported in Chapter Four.

In the present study, performance on tasks that tapped higher level cognitive processes was at best only weakly predicted by the nutritional history variables. There were no relations at all (across villages) between caloric supplementation and scores on verbal inferences, memory for objects, or concept identification. Of the predictors used in our analysis, only SES was a significant predictor of any of these variables. Our results support the speculation that cognitive processes are less vulnerable to the effects of chronic malnutrition than are social-emotional competencies. The evidence from the present study is consistent with previous research. Rutter (1979), reviewing the literature on environmental deprivation and, in particular, maternal deprivation and its effects on cognitive and social-emotional development, noted that disruptions in parent-child relationships that have lasting efffects on later social and emotional development in the child have tended to have insignificant effects on later cognitive functioning (cf. Tizard & Hodges, 1978; Tizard & Rees, 1974). Rutter (1979) suggested further that although the development of social competencies appears to be highly dependent on the learning of social interaction skills in the earliest stages of childhood, and thus on optimal caregiver-child interaction, intellectual development depends more on experiencing some minimal level of interaction with the environment. The implication is that adequate cognitive development may be achieved under less than ideal interpersonal circumstances.

A second line of support comes from a review of 13 studies (Lloyd-Still, 1976) on the effects of malnutrition on cognitive development. Ten of the studies investigated the intellectual development of children over 5 years of age. Several of the studies showed no significant IQ deficits in the previously malnourished group. In all studies, IQ differences between malnourished and control children decreased over time. When malnutrition occurred as a result of neonatal illness but not in the context of poverty or environmental deprivation, there were no lasting effects of malnutrition on intellectual functioning. Lloyd-Still's review suggests that lowered cognitive performance of malnourished children relative to better-nourished controls in studies that report such differences may be related to the general socioeconomic deprivation that characterizes the study populations, of which malnutrition is but one component.

It should be noted that our expectation that malnutrition would be related to attentional characteristics of the child was supported. Although the prediction of individual psychological test performance on the basis of nutritional measures was not as strong as the prediction of the social-emotional outcomes, there generally was consistent prediction of scores on those tests that measured the child's ability to attend to directions, to persist on a difficult task, and to attend to the details of a stimulus configuration. These findings are consistent with the previous research.

While research supports the position that social, emotional, and motivational characteristics may be particularly sensitive indicators of nutritional history, we recognize that a complete separation of social and emotional functioning from cognitive functioning is not possible. The role of affective and motivational factors in cognitive performance is well documented (Zigler and Butterfield, 1968). Cognitive test performance depends on the child's ability to attend, to remain attentive, and to desire to succeed on test items. But, even more important, it is clear that the development of social and affective characteristics intertwines with, rather than runs parallel to, cognitive development (see Sroufe, 1979). For example, it is the child's ability to form a concept of an absent event and to match current experiences to that concept which makes possible the child's sensitivity to separation from parents and wariness of strangers. It is the child's ability to form an expectation (i.e., anticipate an event) that is at the base of the child's capacity for laughter and humor. It is impossible to conceive of such affective developments as a 2-year-old's demandingness and temper tantrums, as well as independence, without the concurrent development of self-awareness. Thus, cognitive developments make possible important social and affective developments. Any attempt to show that only intellectual or only affective characteristics are affected by a particular environmental condition is likely to be fruitless.

What the data *do* show us, however, is that the most sensitive indicators of nutritional insult often are socioemotional, and that it is a particular configuration of behavior impairments which characterizes the undernourished child. The behaviors affected—attention, persistence, distractibility, emotionality, curiosity and exploration, and initiative—may well have repercussions for intellectual test performance and later achievement. The studies cited earlier on chronic undernutrition and cognitive development (Barrett et al., 1981; Edwards & Grossman, 1980; Hicks et al., 1982) suggest such a long-term effect. Further, the effects of malnutrition on these social and emotional behaviors are important for the child's development, whether or not they have a measurable effect on intellectual test performance. Initiative, persistence, control of emotions, and the capacity for sustained involvement are characteristics of the competent child.

In sum, the research gives strong evidence that there are certain behavioral characteristics which are the most sensitive indicators of chronic undernutrition, and that it is the assessment of social and emotional characteristics which may give us the first hint of the child's inadequate health and nutrition.

5.0 CONCLUSIONS AND IMPLICATIONS FOR POLICY AND FUTURE RESEARCH

Chronic undernutrition in the young child has long-term and adverse consequences for development. When a child receives insufficient nutrition, the first aspect of function to be affected is behavior. The child conserves energy, becoming less active and interactive, and as a result loses important opportunities for learning. In the long term, behavior and development are impaired.

It is important to emphasize the environmental context. Critics of research on malnutrition are fond of saying that poverty and deprivation are 'confounding factors' which make it impossible to say that malnutrition was the cause of a child's poor development. The argument appears to be that if, somehow, children could be malnourished in a wholesome environment, there would be no adverse effects. The fact is that malnutrition does occur in the context of poverty and deprivation and not in protective or rehabilitative environments. Most importantly, it is precisely under these conditions of inadequate support and stimulation that malnutrition has maximal adverse effects.

5.1 The Current Concern

Two concerns of policymakers have been with the prevalence of undernutrition in the United States and the extent to which undernutrition is known to be associated with physiological and/or behavioral impairment in childhood and later life. These questions, which were popular topics of social concern at different times in the last three decades, have re-emerged as political issues in the last year. The current concern results largely from the threat of cutbacks in funding for the major federal nutritional programs: food stamps, the school breakfast and lunch programs, and supplemental feeding of women, infants, and children (WIC).

With hunger again emerging as a political issue, there is a renewed debate over the long-term effects of early nutritional deprivation on the child's later intellectual and social development. There is documentation that undernutrition occurring in the context of poverty is linked strongly to maladaptive behavior. It also is recognized, however, that while early and severe malnutrition may affect brain development—in particular, brain cell number and size, myelinization, and dendritic branching—the functional significance of these effects still is being explored. Further, there is evidence that under certain circumstances, nutritional rehabilitation and environmental enrichment may have reparative effects on long-term developmental outcome.

Together, these trends (see Wyman, 1984)—the debate over whether hunger really is a problem in this country, the controversy concerning whether the effects of undernutrition on brain development necessarily impair functioning, and the current tendency to view the effects of even severe malnutrition on cognitive development as being potentially reversible—are leading policymakers to the conclusion that malnutrition is not a major public health problem in the United States. This conclusion is wrong.

We have shown first that undernutrition does exist in the United States, particularly in poorer families. Second, we have shown that there are long-term consequences of chronic undernutrition.

In the research in Guatemala we found a consistent pattern of behavior problems associated with chronic undernutrition: low activity level, little social involvement, infrequent happy expression, poor attention and persistence on problem-solving tasks, and timidity. Studies in Barbados, Jamaica, and South Africa provide supporting results. Over a variety of social circumstances, children with a history of undernutrition are characterized by these behavior impairments.

Would such effects on behavior occur in the United States? While undernutrition is widespread, the magnitude of energy deficit is not as great as in developing countries. Moreover, the environment is less materially impoverished than in the developing world. There are few data directly addressing the issue. But three studies are suggestive. We cited earlier the study in Louisiana by Hicks and colleagues (1982), which examined the effects of the WIC supplementation program on children's cognitive development and behavior. At ages 6 to 10 years, children who had begun participating in the program prenatally and continued for four years showed significantly higher IQ scores and fewer behavior problems than siblings who had begun supplementation after age 1 and continued for three years. In a study of over 7,000 children in the United States, Edwards and Grossman (1980) found that chronic undernutrition, manifested by reduced stature, was a significant predictor of IQ and standardized achievement test scores. The effects were strongest for lower social-economic status children.

Third, in the study in San Diego, we examined the effects of health problems in childhood which resulted in prolonged undernutrition on the child's subsequent cognitive development (Barrett et al., 1981). In the disadvantaged population studied, children whose mothers reported serious health problems for the child during the preschool years which interfered with nutrient intake scored significantly lower at ages 6 to 7 years on standardized tests of intelligence and school achievement than did children whose mothers did not report such problems.

Individual differences in cognitive development reflect, to a large extent, both the biologic integrity of the central nervous system and the child's opportunities for interaction with the physical and social environment. An extended period of undernutrition may result not only in deficits and distortions of the structure of the developing brain but in diminished involvement with the environment and reduced opportunities for learning.

5.2 Future Research Needs

We have come a long way in our understanding of the types of behavior functions influenced by undernutrition. For many years, researchers investigating the functional consequences of undernutrition focused on physiological parameters and disease or on impairments in cognitive functioning as measured by general tests of intelligence. In recent years, however, researchers have become more interested in the effects of malnutrition on children's social and emotional development.

There are several reasons for the shift in emphasis. First, it has become increasingly recognized that it is difficult to interpret data on the effects of malnutrition on cognitive development unless there are data on how malnutrition has affected motivational characteristics. Cognitive test performance is highly dependent on motivational factors including attention, persistence, and self-confidence (see Zigler & Butterfield, 1968). Assessment of motivational and emotional characteristics which may be related to successful test performance is necessary if one is to determine the long-term consequences of undernutrition for intellectual functioning. Second, researchers have begun to reexamine their decision to focus on the cognitive sequelae of human malnutrition to the exclusion of the social and emotional in light of the empirical data on the effects of malnutrition in animals (Barrett, Yarrow, & Klein, 1982; Ricciuti, 1981). The research on malnourished animals identifies an array of behavioral characteristics that are impaired by malnutrition. Some of these behavior dimensions (described in detail in Chapter Three) are activity level, emotional control, attention to novel stimuli, and social responsiveness. The animal research leads us to consider the consequences of malnutrition in humans with respect to social and emotional functioning. Third, as indicated earlier, there is a body of research accumulating which suggests that social and emotional capacities may be more vulnerable to chronic environmental insult than cognitive functions (Rutter, 1979). Longitudinal studies of children deprived of maternal child-rearing and peer experiences (Tizard & Rees, 1974; Tizard & Hodges, 1978) show that generally there are lasting behavior problems as a result of the social deprivation, even when cognitive functioning is not impaired. There is interest in exploring the relative vulnerabilities of cognitive and social functions to different types of social environmental deficits (see Bakeman & Brown, 1980), one of which is chronic undernutrition. Undernutrition is a particularly interesting candidate for consideration as a type of environmental deprivation with social and cognitive as well as biologic consequences because (a) infant feeding occurs within an interactive context, thus constituting a type of social stimulation in and of itself, and (b) interferences with feeding could logically be expected to result in decreased interaction with the physical and social environment, thus resulting in a general lack of environmental stimulation. Finally, the issue of the social and emotional consequences of malnutrition has taken on increased importance with the shift on the part of educators and psychologists from cognitive to social competence as the major criterion for evaluting the psychological functioning of children (Scarr, 1981) and, in particular, as the criterion for studying the effectiveness of intervention programs aimed at improving children's abilities to profit from school and other learning experiences (Zigler & Trickett, 1978).

In summary, there are four reasons for the increased research interest in examining the social and emotional consequences of undernutrition:

1. Recognition of the need to study the processes that may mediate relationships between malnutrition and mental development;

2. Recognition, from animal research, of the probable links between malnutrition and behavioral function;

3. Interest in the issue of social and emotional functions as particularly vulnerable to a range of social-environmental deprivations and biologic insults, including malnutrition; and

4. A general shift in focus from cognitive to social competencies as criteria for evaluting the effectiveness of intervention programs designed to improve children's mental health and psychological well-being.

The concordance of findings on the social and emotional sequelae of chronic malnutrition is powerful. It is clear that when a child receives insufficient food at any period in development, behavior is affected. When the deprivation occurs during infancy, we see immediately behavioral disturbances. The consequences of long-term energy deficit are seen most clearly for those behavior attributes usually described as 'motivational' or affecting 'performance': attentional disturbances, poor persistence, less exploration, and diminished social involvement. The effects of early and sustained food deficit appear to persist through childhood. It seems reasonable to assume that these behavior impairments compromise the development of adult competencies as well.

Given these advances there still are a number of research issues which require attention. These are, first, the relationships between early undernutrition and current nutritional status, including micronutrient deficiencies and the role of the two factors in influencing development; second, neurochemical changes associated with sustained mild to moderate undernutrition and their behavioral correlates; third, the environmental factors which may protect against, compensate for, or reverse the otherwise adverse effects of sustained periods of nutrient deficiency and the processes by which they do so.

5.2.1 Early versus current undernutrition

One of the most difficult problems in interpreting research on the behavioral effects of undernutrition is separating the effects of early undernutrition from those of current nutritional status. In most studies in which early undernutrition has been associated with later behavior impairment, children have lived in a setting in which daily food energy intake continues to be below adequate levels.

In those cases in which malnutrition has occurred in a more protected context (e.g., the studies by Beardslee et al., 1982; Ellis and Hill, 1975; and Klein et al., 1975), adverse effects on behavioral development have generally been attenuated. There is the question of whether differences in results between the two types of studies—those in which malnutrition is accompanied by environmental deprivation and those in which it is not— reflect differences in children's environments at the time of malnutrition, or whether they reflect primarily differences in current nutrition at the time of assessment.

There are resemblances between results of studies that have focused on the long-term sequelae of early malnutrition and studies that have examined the characteristics of acutely malnourished infants prior to or during rehabilitation. Impairments in attention, emotionality and irritability, and generally poor behavior organization are seen in infants with nonorganic failure to thrive (Powell & Low, 1983), children with kwashiokor (Geber & Dean, 1956), and even in adults following prolonged starvation (Keys et al., 1950). Because of the ubiquitousness of this behavior constellation in relation to malnutrition, it is important to determine to what extent the behavior problems represent the concurrent effects of reduced nutrient intake, as opposed to behavioral sequelae of disturbances resulting from long-term deficit.

One study that has attempted to separate acute from chronic components of malnutrition did so statistically and, as reported earlier, found greater evidence for behavioral impairment resulting from long-term deficit than from the current condition. In his study of Ugandan children, Hoorweg (1976) found that the factor representing chronic undernutrition (represented by low weight for age, weight loss after admission, and low hemoglobin level) was significantly related to reasoning ability, arithmetic skills, visual-spatial skills, and memory in school-age children, while an acute factor (represented by serum protein level, edema, skin changes, and a physician's judgment of overall clinical severity) was not.

These factors, however, were derived from scores of malnourished children. That is, the sample consisted exclusively of children who had been hospitalized for acute malnutrition and had recovered to different degrees. Within such a sample, the long-term component might show stronger relationships with behavior outcome than the short-term component. This type of study does not allow inferences about the implications of prior undernutrition in early life relative to current undernutrition at time of assessment in subclinically undernourished children. Behavior impairments in such samples—which represent a large portion of the world's child population—might be more or less related to the extent and duration of early undernutrition than in samples where there was an acute episode of undernutrition resulting in hospitalization.

The findings from the present study suggest a stronger role for early nutritional experience relative to current nutritional status in accounting for the behavior of school-aged children in an endemically malnourished population. Weight for height, a measure of current nutritional status was only weakly related to child behavior, while the child's history of nutritional supplementation up to age 4 years strongly predicted behavior at ages 6 to 8. In addition, there was no evidence that current nutritional status could have been a mediator of the relationship between early nutritional supplementation and later behavior, since weight for height, unlike linear growth, was not predicted by the child's supplementation history.

A related question is the role of micronutrient deficiencies in accounting for the behavioral effects of undernutrition and, in particular, iron deficiency. Iron deficiency is the most prevalent nutritional deficiency of infancy (Oski & McMillan, 1981) and early childhood. According to the Second National Health and Nutrition Examination Survey (Hanes—II; see Guthrie, 1983, p. 573), the mean iron intake for 2- to 3-year-old children was only 62 percent of the standard for age and sex. According to the 1977–1978 Nationwide Food Consumption Survey (Peterkin, 1981), iron deficiency was the most severe nutrient deficiency for both 1- to 2- and 3- to 5-year-olds, with mean percent RDAs of below 70 percent and 70 to 79 percent, respectively.

What makes this issue of theoretical as well as practical importance is that chronic undernutrition often is correlated with iron deficiency (Livingston et al., 1975) so that the child who receives insufficient amounts of food energy during early childhood also is at risk for iron deficiency. Further, the behavior impairments associated with iron deficiency are similar to those observed for chronic energy-protein undernutrition. Oski and McMillan (1981) cited irritability and lack of interest in the environment (p. 161) as typical features of iron deficient infants. In older children, they noted decreased attentiveness, poor behavior control, and distractibility.

Thus, there is a need to study the effects of current micronutrient status as a factor in the relationship between early and severe and/or early and chronic energy-protein undernutrition and later behavior impairment. Unless attention is given to micronutrient parameters, the independent role of the chronic energy-protein component cannot be adequately assessed.

5.2.2 Mechanisms of effects of chronic undernutrition on behavior: Nutrient action on brain function

We have stressed that an important adaptive mechanism in response to inadequate energy intake is decreased physical activity, resulting in lowered energy expenditure. As Viteri (1981) has noted, this has two repercussions: unfavorable effects on social and emotional development, resulting from less interaction with and stimulation from the environment, and retarded growth. If the reduced activity also decreases the efficiency of food utilization and the deposition of lean body mass, as has been shown with malnourished animals and children, the reduction in activity, as well as the deficit of substrate, may have the effect of decreasing the rate of growth.

But the means by which long-term deficits in nutrient intake may have direct, neurochemically mediated effects on behavior are just beginning to be explored. As indicated in Chapters 1 and 3, inadequate nutrition during pregnancy and the first years of life may result in permanent structural changes in fetal and postnatal growth and in the development of the brain (Shoemaker & Bloom, 1977). The structural changes that have been shown to follow severe food deprivation include reduced size and weight of the brain; decreased numbers of neurons and glial cells (depending on the timing and duration of the food restriction); and reduced development of dendritic and axonal processes and synaptic structures (Shoemaker & Bloom, 1977). There is general agreement that growth and structural changes in the brain are most vulnerable to malnutrition occurring during the periods of rapid brain growth (in humans, this would be malnutrition occurring prenatally and in infancy). This would suggest that effects of malnutrition occurring after the first years of life would have less severe consequences for the organism, with respect to neurological development and concomitant behavioral development. Now there are data, however, which suggest that sustained undernutrition at any period during development may have detectable effects on brain biochemistry with possible behavioral sequelae.

Communication between nerves and other cells is mediated neurochemically by chemical compounds called neurotransmitters, most of which are themselves nutrient compounds found in various foodstuffs or are direct metabolic byproducts of dietary components (i.e., certain amino acids). Thus, it is a reasonable hypothesis that the synthesis of some neurotransmitters might be influenced by availability of nutrients involved in their metabolism (Lytle et al., 1984).

The neurotransmitter about which there has been the most research is serotonin. Evidence for nutrition-induced changes in brain serotonergic transmission comes from studies of animals. Rate of synthesis of serotonin appears to depend on the intercellular availability of the precursor amino acid L-tryptophan (Wurtman & Fernstrom, 1976; Wurtman et al., 1980). All tryptophan molecules ultimately derive from the diet. Brain neurons cannot synthesize tryptophan from other constitutents. Thus, to the extent that brain serotonin synthesis is influenced by L-trypotophan availability, adequacy of dietary sources of tryptophan would appear to play a role in serotonergic transmission and its behavioral correlates.

Studies of animals show that malnutrition may result in raising or lowering brain serotonin, depending on the type of nutrient deficit, as well as on the timing and duration of the malnutrition (see Vahvalainen & Oja, 1972).

Although it is difficult to obtain unequivocal evidence that nutrition-induced changes in neurotransmitter biosynthesis have resulted in alterations in behavior (one reason being the difficulty in showing that such alterations in biosynthesis have resulted in changes in the release of neurotransmitters), there are now suggestive data showing effects of short and long-term diet-induced changes in tryptophan on behavior. Relevant to the present issue are data on the effects of chronic malnutrition on animal behavior, apparently mediated by tryptophan deficit. Studies of older animals have shown increased sensitivity to noxious stimuli resulting from reductions in serotonergic transmission due to feeding of a tryptophan-deficient diet. In one study, changes in behavior appeared as soon as low tryptophan-diet rats (corn-fed diet) began consuming the low tryptophan diets, and they continued to be present throughout the duration of the regimen. Behavioral and neurochemical changes were reversible with nutrient-adequate diets (Lytle, et al., 1975).

While the study of the possible impact of short- and long-term dietary insufficiencies on brain neurotransmitters and concurrent behavior impairment is in its infancy, it is reasonable to assume that further research on this topic will suggest that the behavioral impairments we have identified as associated with chronic undernutrition may be influenced to some extent by the effects of short- or long-term unavailability of precursor amino acids on the synthesis of important neurotransmitters. It is important to note that these effects of undernutrition on central nervous system function may occur after central nervous system structure is fully established.

It is interesting to note that behavioral symptoms associated with chronic energy deficit, and in particular distractibility and impulsivity, fit diagnostic criteria for attention deficit disorder (DSM-III, 1982) with or without hyperactivity (Frank, 1984). While attention deficit disorder is assumed to have no unitary cause, it is assumed that central nervous system insult is implicated. Relevant here are findings that depletion of biogenic amine neurotransmitters and, in particular, norepinephrine and dopamine, have been associated with attention deficit disorder (Shaywitz et al., 1978). These data suggest the need for further research on the relationship between early and ongoing protein-energy malnutrition, central nervous system function, and child behavior. This research should extend beyond the postulated critical period for the establishment of brain structure to assess potential ongoing effects on brain function.

5.2.3 Moderating variables: Prevention, compensation, reversibility of effects

A third area involves the question of moderating variables. We have reviewed data showing that there are certain types of experiences that make adverse effects of undernutrition more or less probable, that is, given the same opportunities for nutrient intake. One is the degree to which the mother and/or caregiver recognizes the child's emotional needs, both for attention and stimulation. A second is the overall level of environmental stimulation, represented by global measures of the adequacy of the home or caregiving environment. The research question is: How do these non-nutritional experiences prevent, compensate for, or remedy the effects of malnutrition on behavior? In particular, what are the neural and/or other physiological responses to therapeutic sensory environments?

With regard to mother-child interaction, we have seen that studies of the antecedents of severe malnutrition indicate a role for maternal separation from the child, maternal rejection, and family stress (Chase & Martin, 1970; Geber & Dean, 1956). Studies of mothers of failure-to-thrive (NOFT) children in the United States have shown a correlation of NOFT with maternal depression or psychopathology (Evans et al., 1972; Fischoff et al., 1971; Kerr et al., 1978; Leonard et al., 1966). Such maternal characteristics or experiences have the effect of distorting affective stimuli directed toward the child, and probably tactile stimuli as well.

Zeskind et al. (1984) suggest a probabilistic model for thinking about the effects of poor nutrition for later behavior impairment. He states, "While the malnourished lethargic infant has an increased probability of suboptimal development, a poor developmental outcome is not inevitable since a caregiving environment that is sensitive to and supportive of the infant's needs would have ameliorative effects." (p. 433) The research of Zeskind and Ramey (1978, 1981) provides support. Zeskind and Ramey followed ten fetally malnourished infants of low SES black families from rural North Carolina. All were full-term but classified as prenatally undernourished on the basis of low ponderal index. Five of the infants took part in a day care program beginning at 6 weeks of age. The program involved eight hours a day of social, medical, nutritional, and educational services, five days a week for three years. The control group infants received social, medical, and nutritional services but without the educational curriculum. The authors showed that the children's cognitive and behavioral development was dependent on the quality of the caregiving environment. Bayley developmental quotients were higher in experimental than control infants by 18 months, and differences between the groups were found on the Stanford Binet at 36 months. In addition, the experimental children were less withdrawn and anxious.

Zeskind emphasizes not only the synergism or interaction between nutritional and social-environmental experiences but the fact that the influences are bidirectional. For example, in the intervention study cited above, maternal involvement as well as child behavior was affected by the intervention. Fetally malnourished infants who experienced the intervention elicited greater involvement from mothers—and mothers were not the targets of the intervention; the intervention was infant centered—than did the malnourished infants who did not experience the intervention. Thus, the data suggest that those environmental influences which helped to compensate for the behavioral effects of malnutrition on the child may have done so in part through their indirect effect on the mother; that is, via the child. Chavez's data on maternal and paternal involvement with malnourished children in Mexico also suggest this bidirectional process.

Other clinical interventions have supported the notion that the general level of stimulation of the caregiving environment may play a moderating role with regard to the effects of malnutrition on later behavior. The intervention studies cited earlier by Monckeberg (1981), Grantham-MacGregor (1978, 1979, 1980), and Yaktin and McLaren (1973) all show the importance of psychosocial stimulation, in combination with nutritional and medical care, for the later development of malnourished children.

In addition, there is the fact that adverse effects of malnutrition during infancy on later cognitive development are generally shown when the malnutrition occurs in an impoverished rural environment (see Galler, 1984, for a review) and are less frequently found when the child experiences the malnutrition in a protective, rehabilitative environment; that is, as may occur if a child from a socially privileged family is hospitalized for a disease or operation and undergoes a period of malnutrition secondary to the illness (Lloyd-Still, 1976, gives a review).

But what are the specific processes that account for such moderating effects? Behaviorally, the processes seem clear. We have discussed the models of Lester (1979) and Rossetti-Ferreira (1978), which suggest that malnutrition predisposes the infant through insult to the autonomic and central nervous systems, to become less attentive, responsive, and behaviorally well-organized (e.g., with regard to regulation of state and affect) than the better-nourished infant. These effects on the child make it difficult for the parent to respond sensitively and positively to the child, and the child, in turn, receives even less adequate care and nurturance.

But why should attention from the caregiver—whether touch, visual contact, or auditory stimulation—have a remedial effect on the child's behavioral development? What are the physiological mechanisms?

Studies on the effects of environmental enrichment for animals provide 'leads' but not answers. The research has been conducted primarily on previously deprived animals. The studies have shown that environmental stimulation can lead to increases in brain weight and size (Rosensweig et al., 1972). Formation of synapses also may be environmentally influenced. For example, dark-reared animals showing effects of the light deprivation on formation of synaptic vesicles have shown positive synaptic response to light exposure (Cragg, 1967, 1974). Chemical and electrophysiological recovery (e.g., maturation of visual evoked potentials) also has been demonstrated.

Malnutrition, however, poses a difficult problem. To the extent that malnutrition has resulted in sensory deprivation—and the preceding arguments suggest that even chronic, subclinical malnutrition may result in environmental deprivation—the effects are likely to be nonspecific, not restricted to a particular modality, and inconsistent—at times more severe, at times less. Second, for the study of human malnutrition, it would seem unreasonable, on ethical grounds, to devise tests of therapeutic environments specifically designed to address the effects of a particular presumed deficiency in experience (e.g., tactile stimulation) when a more comprehensive multimodal therapy could be devised. In short, a major question is, How can one carry out the research that will enable us to determine why certain therapeutic environments provided in addition to nutritional resuscitation (i.e., those of Monckeberg, Grantham-McGregor, Zeskind and Ramey, and Yaktin and McLaren) have proved effective in facilitating behavioral and developmental change?

Further, before we can address this issue, and it is a critical one, we first must address the question of *who* is developmentally at risk because of poor nutrition; that is, with whom must one intervene?

Large numbers of children receive insufficient food to sustain normal growth and development. Margen (1984) has suggested that researchers and policymakers may not have the luxury of assuming that all underfed and slowly growing children are developmentally at risk. Most undernourished children are mildly to moderately undernourished with consequent retarded linear growth. Margen contrasts the 'deprivation theory of growth' which is that such growth impairment implies poor health and functional impairment, with the 'homeostatic theory of growth' which states that there is no single optimal potential growth curve for a child but a broad spectrum of growth curves (1984, p. 27). Genetically determined processes provide a 'growth space' for the child, and nutrition and environmental experiences determine the specific growth curve. Thus, different degrees of nutritional adequacy and 'richness' of experience may be compatible with adequate behavioral function.

Margen's model rejects the assumption that one can specify an optimal environment for the child and takes a relativistic non-normative position. "The prevailing theory," he writes, "particularly among nutritionists is that the object of nutrition is to facilitate the 'optimal' genetic potential of each individual. Accordingly, it is assumed that each individual has a certain level of optimal function which can be achieved only in an optimal environment" (p. 25).

"The opposing point of view...is based on a substantially different set of concepts and states that the purpose of the genetic mechanisms is to regulate the functions of an individual to make the best possible adaptation to the environment. Here there can be a vast array of potential responses, depending upon the environmental circumstances. As long as homeostasis can be maintained, the individual will remain in a state of health." (p. 26)

Margen's model is provocative and provides a rationale for studying the question of what degrees of undernutrition and concomitant physiological retardation significantly impair function. It focuses social concern on those who are most nutritionally and environmentally deprived. From our research, however, and the other research reviewed in these chapters, we have seen that subtle changes in behavior often are the sensitive indicators of nutritional history, and the most sensitive indicators of nutritional improvement. In study after study, individual differences in achieved growth and individual differences in nutrient intake *among persons from undernourished populations* have been shown to have important behavioral ramifications. Although the homeostatic theory is appealing in its implication that persons of different cultures and experiences can adapt to the exigencies of their life conditions without being considered 'at risk' or 'abnormal', it is dangerous in its implication that large numbers of persons may adapt to restricted intake of food and specific nutrients without serious consequence. The research suggests that human activity, affect, social relationships, interest in the environment, and opportunities for learning are linked to previous and current nutrient intake and growth. In particular, prospective studies of children recovering from malnutrition suggest that we should be sensitive to the possibility that more, rather than fewer, children who are underfed are missing out on important life opportunities. A major research need is—in continuation of the line of thinking initiated by Margen—to examine the behavioral characteristics of children with known nutritional histories and growth histories so that we can ascertain better the functional relationships—i.e., linear or nonlinear—between nutrition, environmental deprivation, and behavior impairment.

If we can study these questions, we will be in a better position to identify populations of children who are in need of nutritional and social-educational intervention, and to design intervention programs that test our hypotheses about the mechanisms through which social and sensory stimulation can influence behavior function. Such research will have ramifications that will take us beyond the problem of malnutrition and help us to address more broadly the problem of caring for children in need.

References

Aakre, B., Strobel, D.A., Zimmerman, R.R., and Geist, C.R. (1973) Reaction to intrinsic and extrinsic reward in protein malnourished monkeys. *Percept. and Motor Skills* 36:787–790.

Acheson, K.J., Campbell, J.T., Edholm, O.G., Miller, D.S., and Stock, M.J. (1980) The measurement of food and energy intake in man—an evaluation of some techniques. *Am. J. Clin. Nutr.* 33:1147–1154.

Adaic, L.S. and Pollitt, E. (1983) Outcome of maternal nutritional supplementation: A comprehensive review of the Bacon Chow study. *Am. J. Clin. Nutr.* 41:948–978.

Ainsworth, M.D.S. (1979) Infant-mother attachment. *Am. Psychol.* 34:932–937.

———, Blehar, M.C., Waters, E., and Wall, S. (1978) *Patterns of Attachment: A Psychological Study of the Strange Situation.* Hillsdale, NJ: Lawrence Erlbaum.

Als, H., Tronick, E., Adamson, L., and Brazelton, T.B. (1976) The behavior of the full-term but underweight newborn infant. *Dev. Med. Child Neurol.* 18:590–602.

Ashem, B. and Janes, M.D. (1978) Deleterious effects of chronic undernutrition on cognitive abilities. *J. Child Psychol. and Psychiat.* 19:23–31.

Babbie, E.R. (1975) *The Practice of Social Science Research.* Belmont, CA: Wadsworth.

Baird, A., Widdowson, E.M., and Cowley, J.J. (1971) Effects of calorie and protein deficiencies early in life on subsequent learning abilities of rats. *Brit. J. Nutr.* 25:391–403.

Bairagi, R., Chowdhury, M.K., Kim, Y.J., and Curlin, G.T. (1985) Alternative anthropometric indications of mortality. *Am. J. Clin. Nutr.* 42:296–306.

Bakeman, R. and Brown, J.V. (1980) Early interaction: Consequences for social and mental development at three years. *Child Dev.* 51:437–447.

Balasz, R., Lewis, P.T. and Patel, A.J. (1979) Nutritional differences and brain development. In: F. Faulkner and J. Tanner (Eds.), *Human Growth: Neurobiology and Nutrition.* New York: Plenum, pp. 414–472.

Bandura, A., Ross, D., and Ross, S. (1961) Transmission of aggression through imitation of aggressive models. *J. Abnor. and Soc. Psych.* 63:375–382.

Barnes, R.H. (1976) Dual role of environmental deprivation and malnutrition in retarding intellectual development. *Am. J. Clin. Nutr.* 29:912–917.

———, Cunnold, S.R., Zimmerman, R.R., Simmons, H., MacLeod, R.B., and Krook, L. (1966) Influence of nutritional deprivation in early life on learning behaviors of rats as measured by performance in a water maze. *J. Nutr.* 89:399–410.

———, Levitsky, D.A., Pond, W.G., and More, U. (1976) Effects of postnatal dietary protein and energy restriction on exploratory behavior in young pigs. *Dev. Psychobiol.* 9:425–435.

———, Moore, A.U., and Pond, W.G. (1970) Behavioral abnormalities in young pigs caused by malnutrition in early life. *J. Nutr.* 100:149–155.

———, Neely, C.S., Kwong, E., Labadan, B.A., and Frankova, S. (1968) Postnatal nutritional deprivations as determinants of adult rat behavior toward food, its consumption, and utilization. *J. Nutr.* 96:467–476.

Barreda-Moncada, G. (1963) *Estudios sobre alteraciones del crecimiento y del desarrollo Psicologico del Sindrome Pluricarencial (Kwashiorkor).* Caracas, Venezuela: Editora Grafas.

Barrett, D.E. (in preparation) Testing and observing infants and young children: Psychometric issues. In: J.K. Nugent, J. Lydic, and P. Burke, *Newborn and Infant Assessment Techniques.* Boston: Little, Brown.

——— (1987) Undernutrition and child behavior: What behaviors should we measure and how should we measure them? In J. Dobbing (Ed.), *Early Nutrition and Later Achievement.* London: Academic Press, pp. 89–119.

——— (1986a) Behavior as an outcome in nutrition research. *Nutr. Rev. Supp.* 44:224–236.

——— (1986b) Nutrition and social behavior. In: H.M. Fitzgerald, B.M. Lester, and M.W. Yogman (eds.), *Theory and Research in Behavioral Pediatrics, Vol. 3.* New York: Plenum, pp. 147–198.

——— (1984a) Malnutrition and child behavior: Conceptualization, assessment, and an empirical study of social-emotional functioning. In: B. Schurch and J. Brozek (eds.), *Malnutrition and Behavior: A Critical Assessment of Key Issues.* Lausanne, Switzerland: Nestle Foundation for the Study of Nutrition Problems in the World, pp. 280–306.

——— (1984b) Methodological requirements for conceptually valid research studies on the behavioral effects of malnutrition. In: J.R. Galler (ed.), *Nutrition and Behavior.* New York: Plenum, pp. 9–36.

——— (1984c) Methodological issues in nutritional intervention research. In: B. Schurch and J. Brozek (eds.), *Malnutrition and Behavior: A Critical Assessment of Key Issues.* Lausanne, Switzerland: Nestle Foundation for the Study of Nutrition Problems in the World, pp. 585–596.

——— (1984d) The effects of undernutrition on children's behavior. In: D.H. Enstrom (ed.), *Proceedings: Nutrition and Children's Behavior.* Princeton, NJ: Office of Education, Department of Human Services, pp. 1–49.

——— (1982) An approach to the conceptualization and assessment of social-emotional functioning in studying nutrition-behavior relationships. *Am. J. Clin. Nutr.* 35:1222–1227.

——— (1979) A naturalistic study of sex differences in children's aggression. *Merrill-Palmer Quarterly* 134:117–118.

——— and Radke-Yarrow, M.R. (in press) Effects of malnutrition on children's behavior. In: Calloway, D.H. (ed.), *Monograph on the Functional Consequences of Marginal Malnutrition.* Berkeley, CA: Monographs of the Institute of International Studies.

——— and ——— (1985) Effects of nutritional supplementation on children's responses to novel, frustrating, and competitive situations. *Am. J. Clin. Nutr.* 42:102–120.

———, ———, and Klein, R.E. (1982) Chronic malnutrition and child behavior: Effects of early caloric supplementation on social and emotional functioning at school age. *Dev. Psychol.* 18:541–556.

———, ———, Ziegler, M., Livingston, R.B., and Klein, R.E. (1981) Effects of early malnutrition on children's social emotional functioning at school age. Presented at the meeting of the Association for the Care of Children's Health, Toronto, May.

Bartko, J.J. (1966) The intraclass correlation coefficient as a measure of reliability. *Psychol. Repts.* 19:3–11.

Beardslee, W.R., Wolff, P.H., Hurwitz, I., Pankh, R., and Shwachman, H. (1982) The effects of infantile malnutrition on behavioral development: A follow-up study. *Am. J. Clin. Nutr.* 35:437–441.

Beaton, G.H. (1985) Uses and limits of the use of the Recommended Dietary Allowances for evaluating dietary intake data. *Am. J. Clin. Nutr.* 41:155–164.

——— (1983) Energy in human nutrition: Perspectives and problems. *Nutr. Revs.* 41:325–340.

——— (1982) Evaluation of nutrition interventions: Methodologic considerations. *Am. J. Clin. Nutr.* 35:1280–1289.

Behrman, R.E. and Vaughan, V.C. (1983) *Nelson Textbook of Pediatrics.* Philadelphia: W.B. Saunders.

Bengoa, J.M. (1972) Nutritional significance of mortality statistics. In: *Proceedings of the Third Western Hemisphere Nutrition Conference.* New York: Futura, pp. 270–281.

——— (1974) The Problem of Malnutrition. *WHO Chron.* 28:3.

Berwick, D. (1980) Non-organic failure to thrive. *Ped. in Rev.* 1:9.

Bevan, W. and Freeman, O.I. (1952) Some effects of an amino acid deficiency upon the performance of albino rats in a simple maze. *J. Gen. Psychol.* 80:75–82.

Birch, H.G., Pineiro, C., Alcalde, E., Toca, T., and Cravioto, J. (1971) Relation of kwashiorkor in early childhood and intelligence at school age. *Ped. Res.* 5:579–585.

Bithoney, W.G. and Rathbun, G. (1983) Failure to thrive. In: M. Levine, W. Carey, A. Crocker, and R. Gross (eds.), *Developmental Behavioral Pediatrics.* Philadelphia: W.B. Saunders, pp. 557–572.

Bradley, R.H., and Caldwell, B.M. (1984) The relation of infants' home environments to achievement test performance in first grade: A follow-up study. *Child Dev.* 55:803–809.

Brandt, I. (1979) "Growth dynamics of low birthweight infants with emphasis on the perinatal period." In: F. Falkner and J. Tanner (eds.), *Human Growth: Neurobiology and Nutrition.* New York: Plenum Press, pp. 557–617.

Brazelton, T.B., Koslowski, B. and Main M. (1974) The origins of reciprocity: The early mother-infant interaction. In: M. Lewis and L.A. Rosenblum (eds.), *The Effect of the Infant on Its Caregiver.* New York: Wiley.

———, Tronick, E., Lechtig, A., Lasky, R.E., and Klein, R.E. (1977) The behavior of nutritionally deprived Guatemalan infants. *Dev. Med. and Child Neurol.* 19:364–372.

Breslin, R.W., Lonner, W.J., and Thorndike, R.M. (1973) *Cross-cultural Research Methods.* New York: Wiley.

Brockman, L. and Ricciuti, H. (1971) Severe protein-calorie malnutrition and cognitive development in infancy and early childhood. *Dev. Psych.* 4:312–319.

Brozek, J. (1979) Malnutrition and human behavior. In: J. Brozek (ed.), *Behavioral Effects of Energy and Protein Deficits.* Bethesda, MD: Department of

Health, Education, and Welfare (National Institutes of Health).
——— (1978) Nutrition, malnutrition and behavior. *Ann. Rev. Psychol.* 29:157–177.
Butler, R.A. (1965) Investigative behavior. In: A.M. Schrier, H.F. Harlow, and F. Stollnitz (eds.), *Behavior of Nonhuman Primates: Modern Research Trends, Vol. 2.* New York: Academic Press.
Cabak, V. and Najdanvik, R. (1965) Effects of undernutrition in early life on physical and mental development. *Arch. Dis. Child* 40:532–534.
Cairns, R.B. and Green, J.A. (1979) How to assess personality and social patterns: Observations or ratings? In: R.B. Cairns (ed.), *The Analysis of Social Interactions.* Hillsdale, NJ: Lawrence Erlbaum, pp. 209–226.
Caldwell, D.F. and Churchill, J.A. (1967) Learning ability in the progeny of rats administered a protein-deficient diet during the second half of gestation. *Neur.* 17:95–99.
Campbell, D.T. and Frey, P.W. (1970) The implications of learning theory for the fade-out of gains from compensatory education. In: Helmuth, J. (ed.): *Disadvantaged Child, Vol. 3.* New York: Brunner/Mazel, pp. 445–463.
——— and Stanley, J.C. (1966) *Experimental and Quasi-experimental Designs for Research.* Chicago: Rand-McNally.
Celadon, J.M. and de Andraca, I. (1979) Psychomotor development during treatment of severely marasmic infants. *Early Hum. Dev.* 3:267–275.
———, Csaszar, D., Middleton, J., and de Andraca, J. (1980) The effect of treatment on mental and psychomotor development of marasmic infants according to age of admission. *J. Ment. Defic. Res.* 24:27–35.
Champakam, S., Srikantia, S.C., and Gopalan, C. (1968) Kwashiorkor and mental development. *Am. J. Clin. Nutr.* 21:844–852.
Chase, H.P. and Martin, H.P. (1970) Undernutrition and child development. *N. Engl. J. Med.* 282:933–939.
Chavez, A. and Martinez, C. (1979) Consequences of insufficient nutrition on child character and behavior. In: D.A. Levitsky (ed.), *Malnutrition, Environment and Behavior.* Ithaca, New York: Cornell University Press, pp. 238–268.
———, ———, and Yaschine, T. (1975) Nutrition, behavioral development, and mother-child interaction in young, rural children. *Fed. Proceeds.* 34:1574–1582.
Chen, L.C., Chowdhury, A., and Huffman, S.L. (1980) Anthropometric assessment of energy-protein malnutrition and subsequent risk of mortality among preschool aged children. *Am. J. Clin. Nutr.* 33:1836–1845.
Christiansen, N., Vuori, L., Mora, J.O., and Wagner, M. (1974) Social environment as it relates to malnutrition and mental development. In: J. Cravioto, L. Hambraeus, and B. Vahlquist (eds.), *Early Malnutrition and Mental Development.* Uppsala, Sweden: Almquist and Wiksell, pp. 186–199.
Church, C.F. and Church, H.N. (1970) *Food Values of Portions Commonly Used, 11th ed.* Philadelphia: J.B. Lippincott Co.
Cohen, J. (1968) Weighted kappa: Nominal scale agreement with provision for scaled disagreement or partial credit. *Psychol. Bull.* 70:213–220.
Cook, T.D. and Campbell, D.T. (1979) *Quasi-experimentation: Design and Analysis Issues for Field Settings.* Chicago: Rand-McNally.
Coursin, D.B. (1975) Malnutrition, brain development, and behavior: Anatomic, biochemical, and electrophysiologic constructs. In: M.A.B. Brazier (ed.), *Growth and Development of the Brain.* New York: Raven, pp. 289–305.
——— (1972) Nutrition and brain development in infants. *Merrill-Palmer Quarterly* 18:177–202.

Cowley, J.J. and Griesel, R.D. (1966) The effect on growth and behavior of rehabilitating first and second-generation low protein rats. *Anim. Beh.* 14:506–517.

——— and ——— (1964) Low-protein diet and emotionality in the white rat. *J. Genet. Psychol.* 104: 89–98.

——— and ——— (1963) The development of second-generation low-protein rats. *J. Gen. Psych.* 103:233–242.

——— and ——— (1962) Pre and postnatal effects of a low protein diet on the behavior of the white rat. *Psychol. Agric.* 9:216–225.

——— and ——— (1959) Some effects of a low protein diet on first filial generation of white rats. *J. Gen. Psych.* 95:187–201.

Cragg, B.G. (1974) Plasticity of synapses. *Brit. Med. Bull.* 30:141–145.

——— (1967) Changes in visual cortex on first exposure of rats to light: Effect on synaptic dimensions. *Nature* 215:251–253.

Cravioto, J. (1977) Not by bread alone: Effect of early malnutrition and stimuli deprivation on mental development. In: O.P. Ghai (ed.), *Perspectives in Pediatrics.* New Delhi: Interprint, pp. 87–104.

Cravioto, J. and DeLicardie, E. (1972) Environmental correlates of severe clinical malnutrition and language development in survivors from kwashiorkor or malnutrition. In: *Nutrition, the Nervous System, and Behavior.* Pan American Health Organization Scientific Publication No. 251, pp. 73–94.

———, ———, and Birch, H.G. (1966) Nutrition, growth, and neurointegrative development: An experimental and ecologic study. *Ped.* 38(supplement): 319–372.

——— and Robles, B. (1965) Evolution of adaptive and motor behavior during rehabilitation from kwashiorkor. *Am. J. Orthopsych.* 35:449–464.

Crnic, S. (1980) Models of infantile malnutrition in rats: Effects on maternal behavior. *Dev. Psychobiol.* 13:615–628.

Delgado, H., Martorell, R., Brineman, E., and Klein, R. (1982) Nutrition and length of gestation. *Nutr. Res.* 2:117–126.

DeLicardie, E.R. and Cravioto, J. (1974) Behavioral responsiveness of survivors of clinically severe malnutrition to cognitive demands. In: J. Cravioto, L. Hambraeus, and B. Vahlquist (eds.), *Early Malnutrition and Mental Development.* Uppsala, Sweden: Almquist and Wiksell, pp. 134–153.

Denenberg, V.H. (1969) Open-field behavior in the rat: What does it mean? *Ann. N.Y. Acad. Sci.* 159:852–859.

——— (1964) Critical periods, stimulus inputs, and emotional reactivity: A theory of infantile stimulation. *Psychol. Rev.* 71:331–351.

Diagnostic and Statistical Manual of Mental Disorders, Third Edition (DSM-III). (1982) Washington, D.C.: American Psychiatric Association.

DiBenedetta, C. and Cioffa, L.A. (1972) Early malnutrition, brain glycoproteins, and behavior in rats. *Bibl. Nutr. Dieta*, 17:69–82.

Dickerson, J.W.T. (1981) Nutrition, brain growth, and development. In: K.J. Connolly and H.L. Prechtl, eds., *Maturation and Development: Biological and Psychological Perspectives.* Clinics in Developmental Medicine, No. 77/78. Philadelphia: J.B. Lippincott, pp. 110–130.

Dietz, W.H. (1986) Body composition and nutritional assessment of the undernourished child. In: S. Cohen (ed.), *Treating the Underweight Infant, Child and Adolescent.* New York: Appleton, Century Crofts, pp. 1–14.

Dobbing, J. (1985) Infant nutrition and later achievement. *Am. J. Clin. Nutr.* 41:477–484.

——— (1984) Infant nutrition and later achievement. *Nutr. Rev.* 42:1–7.

——— (1974) Prenatal nutrition and neurological development. In: J. Cravioto, L. Hambraeus, and B. Vahlquist (eds.), *Early Malnutrition and Mental Development.* Uppsala, Sweden: Almquist and Wiksell, 96–110.

——— (1964) The influence of early malnutrition on the development of myelination of the brain. *Proc. Roy. Soc. London,* Series B, 159:503–509.

Drillen, C.M., Thomson, A.J.M., and Burgoyne, K. (1980) Low birthweight children at early school-age: A longitudinal study. *Dev. Med. and Child Neurol.*, 22:26–47.

Drotar, D., Nowak, M., Malone, C.A., Eckerle, D. and Negray, J. (1985) Early psychological outcome in failure to thrive: Predictions from an interactional model. *J. Clin. Consul. Psychol.* 14:105–111.

Dubowitz, L.M., Dubowitz, V. and Golderg, D. (1970) Clinical assessment of gestational age in the newborn infant. *J. Ped.* 77:1–10.

Dyson, S.E. and Jones, D.G. (1976) Undernutrition and the developing nervous system. *Prog. in Neurobiol.* 7:171–196.

Edwards, L.N. and Grossman, M. (1980) The relationship between children's health and intellectual development. In: S. Mushkin (ed.), *Health: What Is It Worth? Measures of Health Benefits.* New York: Pergamon Press, pp. 270–313.

Elias, M.F. (1976) Malnutrition in infancy and intellectual development. In: K.F. Reigel and J.A. Meacham (eds.), *The Developing Individual in a Changing World, Vol. 2.* Chicago: Aldine, pp. 454–461.

——— and Samonds, K.W. (1977) Protein and calorie malnutrition: Growth and behavioral development during deprivation and rehabilitation. *Am. J. Clin. Nutr.* 30:355–366.

Ellis, C.E. and Hill, D.E. (1975) Growth, intelligence, and school performance in children with cystic fibrosis who have had an episode of malnutrition during infancy. *J. Ped.* 87:565–568.

Elmer, E. (1960) Failure to thrive: Role of the mother. *Ped.* 25:717–725.

———, Gregg, G.S., and Ellison, P. (1969) Late results of the "Failure to Thrive" syndrome. *Clin. Peds.* 8:584–589.

Engle, P.L., Irwin, M., Klein, R.E., Yarbrough, C., and Townsend, J.W. (1979) Nutrition and mental development in children. In: M. Winick (ed.), *Human Nutrition: A Comprehensive Treatise, Vol. 1.* New York: Plenum, pp. 291–306.

———, Levin, R.J. (1984) Sex differences in the effects of malnutrition on mental development: A review and some hypotheses. In: J. Brozek, B. Schurch (eds.), *Malnutrition and Behavior: Critical Assessment of Key Issues.* Lausanne, Switzerland: Nestle Foundation, pp. 396–410.

———, Yarbrough, C., and Klein, R.E. (1983) Sex differences in the effects of nutrition and social environment on mental development. In: M. Burinic, M.A. Lycette, and W.P. McGreevey (eds.), *Women and Poverty in the Third World.* Baltimore: Johns Hopkins University Press.

Epstein, H.T. (1978) Growth spurts during brain development: Implications for educational policy and practice. In: J.S. Chall and A.F. Mirsky (eds.), *Education and the Brain.* N.S.S.E. Yearbook. Chicago: University of Chicago Press, pp. 343–370.

Epstein, S. (1980) The stability of behavior, II: Implications for psychological research. *Amer. Psychol.* 35:790–808.

Evans, D., Barrie, M.D., Hansen, J.D.L., Moodie, A.D., and van der Spuy, H.I.J. (1980) Intellectual development and nutrition. *J. Ped.* 97:358–363.

———, Moodie, A.D., and Hansen, J.D.L. (1971) Kwashiorkor and intellectual development. *S. Afr. Med. J.* 45:1414–1426.

Evans, S.L., Reinhard, J.B., Succop, R.A. (1972) Failure to thrive: A study of 45 children and their families. *J. Am. Acad. Child Psych.* 22:322–328.

FAO/WHO (1973) Report of a joint FAO/WHO ad hoc expert committee on energy and protein requirements. *FAO Nutr. Meetings Rept. Ser. No. 52, WHO Tech. Rept. Ser. No. 522.*

Fischoff, J., Whitten, C.F., and Petit, M.A. (1971) A psychiatric study of mothers of infants with growth failure secondary to maternal deprivation. *J. Ped.* 79:209–215.

Field, M. (1984) Follow-up developmental status of infants hospitalized for non-organic failure to thrive. *J. Ped. Psychol.* 9:241–256.

Fitch, M.J., Cadol, R.V., Goldson, E., Wendel, T., Swartz, D., and Jackson, E. (1976) Cognitive development of abused and failure to thrive children. *Ped. Psych.* 1:32–37.

Fleischer, S.F. and Turkewitz, G. (1984) The use of animals for understanding the effects of malnutrition on human behavior: Models vs. a comparative analysis. In: J.R. Galler (ed.), *Nutrition and Behavior.* New York: Plenum, pp. 37–61.

——— and ——— (1981) Alteration of maternal behaviors of female rats caring for malnourished pups. *Dev. Psychobiol.* 14:383–388.

Fleiss, J.L. (1981) *Statistical Methods for Rates and Proportions, second edition.* New York: John Wiley.

——— (1973) *Statistical Methods for Rates and Proportions.* New York: John Wiley.

Frank, D.A. (1984) Malnutrition and child behavior: A view from the bedside. In: J. Brozek and B. Schurch (eds.), *Malnutrition and Behavior: Critical Assessment of Key Issues.* Lausanne, Switzerland: Nestle Foundation, pp. 307–326.

Frankova, S. (1981) Influence of early social environment on behavioral development and on later maternal behavior of protein-deprived rats. *Act. Nerv. Super.* 23:81–91.

——— (1974) Interaction between early malnutrition and stimulation in animals. In: J. Cravioto, L. Hambraeus, and B. Vahlquist, eds., *Early Malnutrition and Mental Development.* Uppsala, Sweden: Almqvist and Wikwell, pp. 202–209.

——— (1973a) Influence of the familiarity with the environment and early malnutrition on the avoidance learning and behavior in rats. *Activ. Nerv. Sup.* 15:207–216.

——— (1973b) Effect of protein-calories malnutrition on the development of social behavior in the rat. *Dev. Psychobio.* 6:33–43.

——— and Barnes, R.H. (1968) Effect of malnutrition in early life on avoidance conditioning and behavior of adult rats. *J. Nutr.* 96:485–493.

Freeman, H.E., Klein, R.E., Kagan, J., and Yarbrough, C. (1977) Relations between nutrition and cognition in rural Guatemala. *Am. J. Pub. Health* 67:233–239.

Galler, J.R. (1984) Behavioral consequences of malnutrition in early life. In: J.R. Galler (ed.), *Nutrition and Behavior.* New York: Plenum, pp. 63–118.

———, Ricciuti, H.N., Crawford, M.A. and Kucharski, L.T. (1984) The role of mother-infant interaction in nutritional disorders. In: J.R. Galler (Ed.), *Nutrition and Behavior.* New York: Plenum, pp. 269–304.

———, Ramsey, F., Solimano, G., Lowell, W.E., and Mason, E. (1983a) The influence of early malnutrition on subsequent behavioral development. I. Degree of impairment in intellectual performance. *J. Amer. Acad. Child Psych.* 22:8–15.

———, ———, ———, ———, and ——— (1983b) The influence of early malnutrition on subsequent behavioral development. II. Classroom behavior. *J. Amer. Acad. Child Psych.* 22:16–22.

Gallo, P.V. (1981) Physiological and behavioral consquences of maternal and post-weaning protein restriction. *Phys. and Beh.* 26:77–84.

———, Werboff, J., and Knox, K. (1980) Protein restriction during gestation and lactation: Development of attachment behavior in cats. *Beh. and Neural Bio.* 29:216–223.

Ganz, L. (1968) Generalization behavior in the stimulus deprived organism. In: G. Newton and S. Levin (eds.), *Early Experience and Behavior.* Springfield, IL: Charles C. Thomas, pp. 364–441.

Garcia-Coll, C., Lester, B.M., Valcarcel, M., Hoffman, J. and Oh, W. (1984) Feeding interactions of fetally malnourished infants born to teenage and older mothers. *Pediatric Research* 18(4):96a.

Garfield, R.M. and Rodriguez, P.F. (1985) Health and health services in Central America. *J. Amer. Med. Assoc.* 254:936–943.

Gaston, A.H. (1982) Small-for-gestational age (SGA) infants. In: A.M. Rudolph (ed.), *Pediatrics, 17th ed.* Norwalk, CT: Appleton Century Crofts, pp. 154–157.

Geber, M. and Dean, R.F.A. (1956) The psychological changes accompanying kwashiorkor. *Courrier* 6:3–14.

Gersovitz. M., Madden, J.P., and Smiciklas-Wright, H. (1978) Validity of the 24-hr. dietary recall and seven-day record for group comparisons. *J. Amer. Diet. Assoc.* 73:48–55.

Glaser, H., Heagarty, M., Bullard, D., and Pirchik, E. (1968) Physical and psychological development of children with early failure to thrive. *J. Peds.* 73:5.

Gomez, F., Galvan, R.R., Frenk, S., Munoz, J.C., Chavez, R., and Vazquez, J. (1956) Mortality in second and third-degree malnutrition. *J. Trop. Peds.* 2:77–83.

Gopalan, C. and Ramalingaswami, V. (1955) Kwashiorkor in India. *Ind. J. Med. Res.* 43:751.

Graitcer, P.L. and Gentry, E.M. (1981) Measuring children: One reference for all. *The Lancet* 2:297–299.

Grantham-McGregor, S. (1984a) The social development of an adopted child recovering from malnutrition. *Nutr. Revs.* 42:16–19.

——— (1984b) Chronic undernutrition and cognitive abilities. *Hum. Nutr: Clin.* 38:83–94.

——— (1984c) Rehabilitation following clinical malnutrition. In: J. Brozek and B. Schurch (eds.), *Malnutrition and Behavior: Critical Assessment of Key Issues.* Lausanne, Switzerland: Nestle Foundation, pp. 531–554.

———, Schofield, W., and Harris, L. (1983) The effect of long-term psychosocial stimulation on mental development of severely malnourished children. *Peds.* 72:239–243.

———, Stewart, M., and Desai, P. (1978) A new look at the assessment of mental development of severely malnourished children. *Dev. Med. Child Neurol.* 20:239–243

———, ———, Powell, C., and Schofield, W.N. (1979) Effect of stimulation on mental development of malnourished child. *Lancet* 2:200–201.

———, ———, and Schofield, W.N. (1980) Effect of long-term psychological stimulation on mental development of severely malnourished children. *Lancet* 2:785–789.

Graves, P.L. (1984) Infant undernutrition in a Nepali village: Behaviors and beliefs. *J. Trop. Ped.* 30:176–183.

——— (1978) Nutrition and infant behavior: A replication study in the Katmandu Valley, Nepal. *Am. J. Clin. Nutr.* 31:541–551.

——— (1976) Nutrition, infant behavior, and maternal characteristics: A pilot study in West Bengal, India. *Am. J. Clin. Nut.* 29:305–319.

Greenough, W.T. (1975) Experimental modification of the developing brain. *Am. Sci.* 63:37–46.

Gurney, J.M. (1979) The young child: Protein energy malnutrition. In: D.B. Jelliffe and E.F.P. Jelliffe (eds.), *Nutrition and Growth.* New York: Plenum.

Guthrie, H.A. (1983) *Introductory Nutrition, Fifth Edition.* St. Louis: Mosby.

——— (1968) Severe undernutrition in early infancy and behavior in rehabilitated albino rats. *Phys. and Beh.* 3:619–623.

Habicht, J.P., Martorell, R., Yarbrough, C., Malina, R.M., and Klein, R.E. (1974) Height and weight standards for preschool children: How relevant are ethnic differences in growth potential? *Lancet* 611–614.

———, Yarbrough, C., and Klein, R.E. (1974) Assessing nutritional status in a field study of malnutrition and mental development. In: J. Cravioto, L. Hambraeus, and B. Vahlquist, (eds.), *Methodology in Studies of Early Malnutrition and Mental Development.* Uppsala, Sweden: Almqvist and Wiksell, pp. 35–42.

———, ———, Lechtig, A., and Klein, R.E. (1976) Relationships of birthweight, maternal nutrition and infant mortality. *Nut. Rep. Int.* 7:533–546.

Halas, E.S. and Sandstead, H.H. (1980) Malnutrition and behavior: The performance versus learning problem revisited. *J. Nutr.* 110:1858–1864.

Hamill, P.V.V., Drizd, T.A., Johnson, C.L., Reed, R.B., Roche, A.F. and Moore, W.M. (1979) Physical growth: National Center For Health Statistics percentages. *Am. J. Clin. Nutr.* 32:607–629.

Hanson, H.M. and Simonson, M. (1971) Effects of fetal undernourishment on experimental anxiety. *Nutr. Repts. Inter.* 4:307–314.

Hansen, J.P.L., Freeman, C., Moodie, A.D., and Evans, D.E. (1971) What does nutritional growth retardation mean? *Pediatrics* 47:299–313.

Harlow, H.F. (1965) Total social isolation in monkeys. *Proc. Nat. Acad. Scien.* 59:90–97.

———, Harlow, M.K., Schiltz, K.A., and Mohr, M.K. (1968) The effects of early adverse and enriched environments on the learning ability of rhesus monkeys. In: L.E. Jarrond (ed.), *Cognitive Processes of Nonhuman Primates.* New York: Academic Press, pp. 121–148.

Hebb, D.O. and Williams, L. (1946) A method of rating animal intelligence. *J. Gen. Psych.* 34:59–65.

Heimendinger, J. and Laird, N. (1983) Growth changes: Measuring the effect of an intervention. *Eval. Rev.* 7:80–95.

Herrera, M.G., Mora, J.O., Christiansen, N., Ortiz, N., Clement, J., Vuori, L., Waber, D., DeParedes, B., and Wagner, M. (1980) Effects of nutritional supplementation and early education on physical and cognitive development. In: R.R. Turner and F. Reese (eds.), *Life-span Developmental Psychology.* New York: Academic Press, pp. 149–184.

Hertzig, M.E., Birch, H.G., Richardson, S.A., and Tizard, J. (1972) Intellectual levels of school children severely malnourished during the first two years of life. *Ped.* 49:814–823.

Hicks, L.E., Langham, R.A., and Takenaka, J. (1982) Cognitive and health measures following early nutritional supplementation: A sibling study. *Am. J. Public Health* 72:1110–1118.

Hiimes, J.H., Roche, A.F., Thissen, D. and Moore, W.M. (1985) Parent specific adjustments for evaluation of recumbent length and stature of children. *Pediatrics* 75:304–313.

Hoorweg, J. (1976) *Protein-energy Malnutrition and Intellectual Abilities.* The Hague: Mouton.

———, Stanfeld, J. (976) The effects of protein and energy malnutrition in early childhood on intellectual and motor activities in later childhood and adolescence. *Dev. Med. Child Neurol.* 18:330–350.

Horton, E.S. (1982) Effects of low energy diets on work performance. *Am. J. Clin. Nutr.* 35:1228–1223.

Hubbard, V.S. and Hubbard, L.R. (1985) Clinical assessment of nutritonal status. In: W.A. Walker and J.B. Watkins (eds.), *Nutrition in Pediatrics: Basic Science and Clinical Application.* Boston: Little, Brown, pp. 121–150.

Hufton, I.W. and Oates, R.K. (1977) Nonorganic failure to thrive: A long-term follow-up. *Ped.* 59:73–79.

Im, H.S., Barnes, R.H., and Levitsky, D.A. (1976) Effect of early protein-energy malnutrition and environmental changes on cholinesterase activity of brain and adrenal glands of rats. *J. Nutr.* 106: 342–349.

Jelliffe, D.B. (1966) Assessment of nutritional status of the community. *Mon. Vol. 53.*. Geneva: WHO.

——— (1965) Effect of malnutrition on behavioral and social development. In: *Proceedings of Western Hemisphere Nutrition Congress.* Chicago: American Medical Association.

Jensen, A.R. (1974) Cumulative deficit: A testable hypothesis? *Dev. Psychol.* 10:996–1019.

——— (1977) Cumulative deficit of IQ in blacks in the rural South. *Dev. Psychol.* 13:184–191.

Johnson, S.M. and Bolstad, O.D. (1973) Methodological issues in naturalistic observation: Some problems and solutions for field research. In: L.A. Hamerlynck, L.C. Handy, and E.J. Mash, eds., *Behavior Change: Methodology, Concepts, and Practice.* Champaign, IL: Research Press, pp. 1–67.

Johnston, E.F. and Lampl, M. (1984) Anthropometry in studies of malnutrition and behavior. In: J. Brozek and B. Schurch (eds.), *Malnutrition and Behavior: Critical Assessment of Key Issues.* Lausanne, Switzerland: Nestle Foundation, pp. 51–70.

Joos, S.K., Pollitt, E., Mueller, W.H., and Albright, D.L. (1983) The Bacon-Chow study: Maternal nutritional supplementation and infant behavioral development. *Child Dev.* 54:669–676.

Karlberg, P., Engstrom, I., Karlberg, J., and Kristiansson, B. (1980) Evaluation of growth during the first two years of life. In: B. Kristiansson (ed.), *Low Rate of Weight Gain in Infancy and Early Childhood.* Goteborg, Sweden: Department of Paediatrics, University of Goteborg, pp. 35–47.

Keen, D.V. and Pearse, R.G. (1985) Birthweight between 14 and 42 weeks gestation. *Archs. Dis. in Child.* 60:440–446.

Kerr, M.A.D., Bognes, J.L., and Kerr, D.S. (1978) Psychosocial functioning of mothers of malnourished children. *Ped.* 62:778–784.

Keys, A., Brozek, J., Henschel, A., Mickelson, O., and Taylor, H.G. (1950) *The Biology of Human Starvation, Vol. II.* Minneapolis, MN: University of Minnesota Press.

Kielmann, A.A., and McCord, C. (1978) Weight-for-age as an index of risk of death in children. *Lancet* i:1247–1250.

Kimble, G.A. (ed.) (1961) Hilgard and Marquis' *Conditioning and Learning.* New York: Appleton-Century-Crofts.

Kimble, K.J., Ariagno, R.L., Stevenson, D.K., and Sunshine, P. (1982) Growth to age 3 years among very low birthweight sequalae-free survivors of modern neonatal intensive care. *J. Peds.* 100:622–624.

Klein, P.S., Forbes, G.B., and Nader, P.R. (1975) Effects of starvation in infancy (pyloric stenosis) on subsequent learning abilities. *J. Pediatr.* 87:8–15.

Klein, R.E. (1979) Malnutrition and human behavior: A backward glance at an ongoing longitudinal study. In: D.A. Levitsky (ed.), *Malnutrition, Environment, and Behavior.* Ithaca, NY: Cornell University Press, pp. 219–237.

———, Irwin, M., Engle, P.L., and Yarbrough, C. (1977) Malnutrition and mental development in rural Guatemala. In: N. Warren (ed.), *Studies in Cross-Cultural Psychology.* New York: Academic Press, pp. 91–119.

———, Yarbrough, C., Lasky, R.E., and Habicht, J.P. (1974) Correlations of mild to moderate protein-calorie malnutrition among rural Guatemalan infants and preschool children. In: J. Cravioto, L. Hambraeus, and B. Vahlquist (eds.), *Early Malnutrition and Mental Development.* Uppsala, Sweden: Almquist and Wiksell, pp. 168–181.

Kotelchuck, M. (1980) Non-organic failure to thrive: The status of interactional and environmental etiologic theories. In: B. Camp (ed.), *Advances in Behavioral Pediatrics.* New York: Academic Press.

Krantzler, N.J., Mullen, B.J., Comstock, E.M., Holden, C.A., Schutz, H.G., Grivetti, L.E., and Meiselman, H.L. (1982a) Methods of food intake assessment—an annotated bibliography. *J. Nutr. Educ.* 14:108–119.

———, ———, Schutz, H.G., Grivetti, L.E., Holden, C.A., and Meiselman, H.L. (1982b) Validity of telephoned diet recalls and records for assessment of individual food intake. *Am. J. Clin. Nutr.* 36:1234–1242.

Latham, M.C. (1974) Protein-calorie malnutrition in children and its relation to psychological development and behavior. *Phys. Revs.* 59:541–565.

——— (1968) Pre- and postkwashiorkor and marasmus. In: R.A. McCance and E.M. Widdowson (eds.), *Protein Deficiencies and Calorie Deficiencies.* London: Churchill.

Lavigne, J. and Ryan, M. (1979) Psychologic adjustment of siblings of children with chronic illness. *Pediatrics* 63:616–627.

Lechtig, A., Delgado, H., Lasky, R., Yarbrough, C., Martorell, R., Habicht, J.P., and Klein, R.E. (1975a) Effects of maternal nutrition on fetal growth and infant development. *Pan Am Health Org. Bull.* 10:301–316.

———, Martorell, R., Yarbrough, C., Delgado, H. and Klein, R.E. (1976) The urea/creatinine ratio: Is it useful for field studies? *J. Trop. Pediatr.* 22:121–128.

———, Yarbrough, C., Delgado, H., Habicht, J.P., Martorell, R., and Klein, R.E. (1975b) Effects of moderate maternal malnutrition on birthweight. *Am. J. Clin. Nutr.* 28: 1223–1233.

Leibel, R.L., Pollitt, E., Kim, I., and Viteri, F. (1982) Studies regarding the impact of micronutrient status on behavior in man: Iron deficiency as a model. *Am. J. Clin. Nutr.* 35:1211–1221.

Lenneberg, E.H. (1969) On explaining language. *Science* 164:635–643.

Leonard, M.F., Rhymes, J.P., and Solnit, A.J. (1966) Failure to thrive in infants: A family problem. *Am. J. Dis. Child* iii:600–612.

Lester, B.M. (1979) A synergistic process approach to the study of prenatal malnutrition. *Inter. J. Beh. Dev.* 2:377–394.

——— (1976) Spectrum analysis of the cry sounds of well-nourished and malnourished infants. *Child Dev.* 47:237–241.

——— (1975) Cardiac habituation of the orienting response to an auditory signal in infants of varying nutritional status. *Dev. Psychol.* 11:432–442.

———, Garcia Coll, C., Valcarcel, M., Hoffman, J., and Brazelton, T.B. (1986) Effects of atypical patterns of fetal growth on newborn (NBAS) behavior. *Child Dev.* 57:11–19.

——— and Zeskind, P.S. (1982) A biobehavioral perspective on crying in early infancy. In: H. Fitzgerald, B. Lester, and M. Yogman (eds.), *Theory and Research in Behavioral Pediatrics, Vol. I.* New York: Plenum, pp. 133–180.

Levine, S., Chevalier, J.A., and Korchin, S.J. (1956) The effects of shock and handling in infancy on later avoidance learning. *J. Person.* 24:475–493.

——— and Wiener, S.H. (1976a) Malnutrition and early environmental experience: Possible interaction effects on later behavior. In: R.N. Walsh and W.T. Greenough (eds.), *Environments as Therapy for Brain Dysfunction.* London: Plenum, pp. 51–70.

——— and ——— (1976b) A critical analysis of data on malnutrition and behavioral deficits. *Advs. in Ped.* 22:113–136.

Levitsky, D.A. (1979) Malnutrition and the hunger to learn. In: D.A. Levitsky (ed.), *Malnutrition, Environment, and Behavior.* Ithaca, NY: Cornell University Press, pp. 161–179.

——— and Barnes, R.H. (1975) Malnutrition and the biology of experience. In: A. Chavez, H. Bourges, and S.S. Basta (eds.), *Proceedings of the IXth International Congress on Nutrition.* Basel: S. Karger, pp. 330–334.

——— and ——— (1972) Nutritional and environmental interactions in the behavioral development of the rat: Long-term effects. *Science* 176: 68–71.

——— and ——— (1970) Effect of early malnutrition on the reaction of adult rats to aversive stimuli. *Nature* 225:468–469.

———, Goldberger, L., and Massaro, T.F. (1979) Malnutrition, learning, and animal models of cognition. In: M. Winick (ed.), *Nutrition, Pre-, and Postnatal Development.* New York: Plenum, pp. 273–290.

——— and Strupp, B.J. (1984) Functional isolation in rats. In: J. Brozek and B. Schurch (eds.), *Malnutrition and Behavior: Critical Assessment of Key Issues.* Lausanne, Switzerland: Nestle Foundation, pp. 411–420.

Lewis, M., Feiring, C., McGuffog, C., and Jaskir, J. (1984) Predicting psychopathology in six-year-olds from early social relations. *Child Dev.* 55:123–136.

Liebert, R.M. and Baron, R.A. (1972) Some immediate effects of televised violence on children's behavior. *Dev. Psychol.* 6:469–475.

Livingston, R.N., Calloway, D.H., MacGregor, J.S., Fisher, G.J., and Hastings, A.B. (1975) U.S. poverty impact on brain development. In: M.A.B. Brazier (ed.), *Growth and Development of the Brain.* New York: Raven Press, pp. 377–394.

Lloyd-Still, J.D. (1976) Clinical studies on the effects of malnutrition during infancy on subsequent physical and intellectual development. In: J.D. Lloyd-Still (ed.), *Malnutrition and Mental Development.* Littleton, MA: Publishing

Sciences Group, pp. 103–161.

———, Hurwitz, I., Wolff, P., and Scwachman, H. (1974) Intellectual development after severe malnutrition in infancy. *Pediatrics* 54:306–311.

——— and Schwachman, N. (1974) Intelligence after malnutrition. *Lancet* 1:679.

Lorenz, K. (1957) Companionship in bird life. In: C. Scholler (ed.), *Instinctive Behavior.* New York: International Universities Press.

Lozoff, B., Wolf, A.W., Urrutia, J.J., and Viteri, F.E. (1985) Abnormal behavior and low developmental test scores in iron-deficient anemic infants. *J. Dev. Behav. Ped.* 6:69–75.

Lubchenco, L.O., Hansman, C., Dressler, M., and Boyd, E. (1963) Intrauterine growth as estimated from liveborn birth-weight data at 24 to 42 weeks of gestation. *Ped.* 32:793–800.

Lytle, L.D., Messing, R.B., Fisher, L., and Phebus, L. (1975) Effects of long-term corn consumption on brain serotonin and the response to electric shock. *Science* 190:692–694.

———, Whitacre, C.S., and Nelson, M.F. (1984) Mechanisms of nutrient action on brain function. In: J.R. Galler (ed.), *Nutrition and Behavior.* New York: Plenum, pp. 223–265.

Madden, J.P., Goodman, S.J., and Guthrie, H.A. (1976) Validity of the 24 hour recall. *J. Am. Diet. Assoc.* 68:143–144.

Magnussen, D. (1966) *Test Theory.* Reading, MA: Addison-Wesley.

Mandler, J.M. (1958) Effects of early food deprivation on adult behavior in the rat. *J. Comp. and Physiol. Psych.* 51:513–517.

Margen, S. (1984) Energy-protein malnutrition: The web of causes and consequences. In: J. Brozek and B. Schurch (eds.), *Malnutrition and Behavior: Critical Assessment of Key Issues.* Lausanne, Switzerland: Nestle Foundation, pp. 20–31.

Martorell, R. (1980) Interrelationships between diet, infectious disease, and nutritional status. In: L.S. Green and F.E. Johnston (eds.), *Social and Biological Predictors of Nutritional Status, Physical Growth, and Neurological Development.* New York: Academic Press, pp. 81–108.

———, Lechtig, A., Yarbrough, C., Delgado, H., and Klein, R.E. (1976) Protein-calorie supplementation and postnatal physical growth: A review of findings from developing countries. *Achivos Latinoamer. de Nutr.* 26:116–128.

Mason, W. and Green, P.H. (1962) The effects of social restriction on the behavior of rhesus monkeys. IV. Responses to novel environment and to an alien species. *J. Comp. Physiol. Psychol.*, 55:363–368.

The 1983 Massachusetts Nutrition Survey (*MNS*). (1983). Boston: Massachusetts Department of Public Health.

McKay, H., Sinesterra, L., McKay, A., Gomez, H., and Lloreda, P. (1978) Improving cognitive ability in chronically deprived children. *Science* 200: 270–278.

McKelvey, R.K. and Marx, M.H. (1951) Effects of infantile food and water deprivation on adult hoarding in the rat. *J. Comp. and Physiol. Psych.* 44:423–430.

McLaren, D.S., and Read, W.W.C. (1975) Weight/length classification of nutritional status. *Lancet* ii:219–221.

———, Yaktin, U.S., Kanawati, A., Sabbagh, S., and Kadi, Z. (1973) The subsequent mental and physical development of rehabilitated marasmic infants. *J. Mental Deficiency Res.* 17:273–281.

Miller, H.C. and Jekel, J.F. (1985) Diagnosing intrauterine growth retardation in newborn infants. *Perinatology/Neonatology* Sept./Oct.:35–42.

———, Hassanein, K. (1971) Diagnosis of impaired fetal growth in newborn infants. *Ped.* 48:511–522.

Mitchell, G.D. and Clark, D.L. (1968) Long-term effects of social isolation in nonsocially adapted rhesus monkeys. *J. Genet. Psychol.* 113:117–128.

Mitchell, W.G., Gorell, R.W., and Greenberg, R.A. (1980) Failure to thrive: A study in a primary care setting, epidemiology and follow-up. *Ped.* 65:971–977.

Monckeberg, F. (1981) Treatment of severe malnutrition during the first year of life. In: N. Selvey and P.L. White (eds.), *Nutrition in the 1980's: Constraints on Our Knowledge*. New York: Alan R. Liss, pp. 141–152.

——— (1979) Recovery of severely malnourished: Effect of early sensory-affective stimulation. In: J. Brozek (ed.), *Behavioral Effects of Energy and Protein Deficits*. Bethesda, MD: DHEW, pp. 121–130.

——— (1968) Effect of early marasmic malnutrition on subsequent physical and psychological development. In: N.S. Scrimshaw and J.E. Gordon (eds.), *Malnutrition, Learning and Behavior.* Cambridge, MA: MIT Press, pp. 269–278.

Mook, D. (1983) In defense of external invalidity. *Am. Psychol.* 38:379–387.

Mora, J.O. (1984) Nutritional assessment by anthropometry: Prevalence studies. In: J. Brozek and B. Schurch (eds.), *Malnutrition and Behavior: Critical Assessment of Key Issues*. Lausanne, Switzerland: Nestle Foundation, pp. 98–106.

———, Amezquita, A., Castro, L., Christiansen, N., Clement-Murphy, J., Cobos, L., Cremer, H., Dragastin, S., Elias, M., Franklin, D., Herrera, M., Ortiz, N., Pardo, R., de Paredes, B., Ramos, C., Riley, R., Rodriguez, H., Vuori-Christiansen, L., Wagner, M., and Stare, F.J. (1974) Nutrition, health and social factors related to intellectual perormance. *World Rev. Nutr. Diet* 19:205–236.

———, Clement, J., Christiansen, N., Ortiz, N., Vuori, L., and Wagner, M. (1979) Nutritional supplementation, early stimulation, and child development. In: J. Brozek (ed.), Behavioral Effects of Energy and Protein Deficits. Washington, D.C.: DHEW, pp. 255–269.

Morris, C.J. (1974) The effects of early malnutrition on one-way and two-way avoidance behavior. *Physiol. Psych.* 2:148.

Morrison, H.L. and McKinney, W.T., Jr. (1976) Environments of dysfunction: The relevance of primate animal models. In: R.N. Walsh and W.T. Greenough (eds.), *Environments as Therapy for Brain Dysfunction.* London: Plenum, pp. 132–170.

Myslivecek, J., Hassmannova, J., and Springer, V. (1969) Conditioning (avoidance reaction) in rats with different individual development. *Activ. Nerv. Sup.* 11:280–282.

Naeye, R.L. (1979) Weight gain and the outcome of pregnancy. *Am. J. Obstet. Gynecol.* 135:3–9.

National Center for Health Statistics (1976) *NCHS Growth Charts*. Monthly Vital Statistics Report, Vol. 25, No. 3, Supp. ((HRA) 76–1120.) Rockville, MD: Health Resources Administration, June.

National Research Council (1980) *Recommended Daily Allowances.* (Ninth rev. ed.) Washington, D.C.: National Academy of Sciences.

Nerlove, S.B., Roberts, J.M., Klein, R.E., Yarbrough, C., and Habicht, J.P. (1974) Natural indicators of cognitive development: An observational study of rural Guatemalan children. *Ethos* 2:265–295.

Neumann, C.G., Jelliffe, D.B., and Jelliffe, J. (1978) Interaction of nutrition and infection. *Clin. Ped.* 11:807–811.

Nie, N.H., Hull, C.H., Jenkins, J.G., Steinbrenner, K., and Bent, D.H. *Statistical Package for the Social Sciences* (2nd ed.). New York: McGraw-Hill.

Oates, R.K., Peacock, A., and Forrest, D. (1985) Long-term effects of nonorganic failure to thrive. *Ped.* 75:36–40.

Oski, F.A. and Honig, A.S. (1978) The effects of therapy on the developmental scores of iron-deficient infants. *J. Ped.* 92:21–25.

——— and McMillan, J.A. (1981) Iron in infant nutrition. In: R.M. Suskind (ed.), *Textbook of Pediatric Nutrition.* New York: Raven, pp. 153–162.

Ottinger, D.R. and Tanabe, G. (1969) Maternal food restriction: Effects on offspring behavior and development. *Dev. Psychobio.* 2:7–9.

Owen, G.M., Kram, R.M., Garry, P.J., Lowe, J.E., and Lubin, A.H. (1974) A study of nutritional status of preschool children in the United States, 1968–1970. *Pediatrics* (suppl.) 53(Part 2):597–646.

——— and Owen, A.L. (1982) Nutritional status of North Americans. In: A. Neuberger and T.H. Jukes (eds.), *Human Nutrition: Current Issues and Controversies.* Englewood, NJ: Burgess, pp. 73–86.

Patton, R.G. and Gardner, L.I. (1962) Influence of family environment on growth: The syndrome of "maternal deprivation." *Ped.* 30:957–962.

Payne, F.D. (1980) Children's prosocial conduct in structured situations and as viewed by others: Consistency, convergence, and relationships with person variables. *Child Dev.* 51:1252–1259.

Peterkin, B.B. (1981) Nationwide food consumption survey, 1977–1978. In: N. Selvey and P.L. White (eds.), *Nutrition in the 1980's: Constraints on Our Knowledge.* New York: Alan R. Liss, pp. 59–69.

Piaget, J. (1970) Piaget's theory. In: P.H. Mussen (ed.), *Carmichael's Manual of Child Psychology, Vol I.* New York: Wiley, pp. 703–732.

Picone, T.A., Allen, L.H., Olsen, P.N., and Ferris, M.E. (1982a) Pregnancy outcome in North American women. II. Effects of diet, cigarette smoking, stress and weight gain on placentas, and on neonatal physical and behavioral characteristics. *Am. J. Clin. Nutr.* 36:1214–1224.

———, ———, Schramm, M.M., and Olsen, P.N. (1982b) Pregnancy outcome in North American women. I. Effects of diet, cigarette smoking, and psychological stress on maternal weight gain. *Am. J. Clin. Nutr.* 36:1205–1213.

Platt, B.S. and Stewart, R.J.C. (1968) Effects of protein-calorie deficiency on dogs. Vol. 1: Reproduction, growth and behavior. *Dev. Med. and Child Neurol.* 10:3–24.

Plomin, R., DeFries, J.C., and Loehlin, J.C. (1977) Genotype-environment interaction and correlation in the analysis of human behavior. *Psych. Bull.* 84:309–322.

Pollitt, E. (1980) *Poverty and Malnutrition in Latin America.* New York: Praeger.

——— (1975) Failure to thrive: Socioeconomic, dietary intake, and mother-child interaction data. *Fed. Proc.* 34:1593–1597.

——— and Eichler, A. (1976) Behavioral disturbances among failure to thrive children. *Am. J. Dis. Children* 130:24–29.

——— and Granoff, D. (1967) Mental and motor development of Peruvian children treated for severe malnutrition. *Rev. Interam. Psychol.* 1:93–102.

——— and Thomson, C. (1977) Protein-calorie malnutrition and behavior: A view from psychology. In: R.J. Wurtman and J.J. Wurtman (eds.), *Nutrition and the Brain, Vol. 2.* New York: Raven, pp. 261–306.

——— and Wirtz, S. (1981) Mother-infant feeding interaction and weight gain in the first month of life. *J. Amer. Diet. Assoc.* 78:596–600.

Powell, G.F., Brasel, J.A., and Blizzard, R.M. (1967) Emotional deprivation and growth retardation simulating idiopathic hypopituitarism. *New Eng. J. Med.* 276:1271–1276.

——— and Low, J. (1983) Behavior in nonorganic failure to thrive. *Dev. and Beh. Ped.* 4:26–33.

Preliminary Findings of the First Health and Nutrition Survey (1975). Washington, D.C.: DHEW (DHEW Pub. No. (HRA)74–1219).

Prentice, A.M., Roberts, S.B., Watkinson, M., Whitehead, R.G., Paul, A.A., Prentice, A., and Watkinson, A.A. (1980) Dietary supplementation of Gambian nursing mothers and lactational performance. *Lancet* ii(8200): 886–887.

Quinton, D. and Rutter, M. (1976) Early hospital admissions and later disturbances of behavior. *Dev. Med. Child Neurol.* 18:447–459.

Rabinowitz, M.S. and Rosvale, H.E. (1951) A closed field intelligence test for rats. *Can. J. Psych.* 5:122–128.

Rajalakshmi, R. (1975) Factors affecting psychological development in man. In: M.A.B. Brazier (ed.), *Growth and Development of the Brain.* New York: Raven, pp. 355–375.

———, Govindarajan, K.R., and Ramakrishnan, C.U. (1965) Effect of dietary protein content on visual discrimination learning and brain biochemistry in the albino rat. *J. Neurochem.* 12:261–271.

Randt, C.T. and Derby, B.M. (1973) Behavioral and brain correlates in early life nutritional deprivation. *Arch. Neurol.* 28:167–172.

Reed, R.B. and Stuart, H.C. (1959) Patterns of growth in height and weight from birth to eighteen years of age. *Pediatrics* 24:904–921.

Ricciuti, H.N. (1981) Developmental consequences of malnutrition in early childhood. In: M.Lewis and L. Rosenblum (eds.), *The Uncommon Child.* New York: Plenum, pp. 150–172.

——— (1974) Assessing the interaction of nutritional and socio-environmental influences on development. In: J. Cravioto, L. Hambraeus, and B. Vahlquist (eds.), *Methodology in Studies of Early Malnutrition and Mental Development.* Sweden: Almquist and Wiksell, p. 182–185.

Richardson, S.A. (1979) Severity of malnutrition in infancy and its relation to later intelligence. In: J. Brozek (ed.), *Behavioral Effects of Energy and Protein Deficits.* Bethesda, MD: DHEW, pp. 172–184.

——— (1976) The influence of severe malnutrition in infancy on the intelligence of children at school age: An ecological perspective. In: R.N. Walsh and W.T. Greenough (eds.), *Environments as Therapy for Brain Dysfunction.* London: Plenum, pp. 256–275.

——— (1974) The background histories of schoolchildren severely malnourished in infancy. *Advs. in Ped.* 21:167–195.

——— (1972) Ecology of malnutrition: Nonnutritional factors influencing intellectual and behavioral development. In: *Nutrition, the Nervous System and Behavior.* Pan American Health Organization, Scientific Publication No. 251, pp. 101–110.

———, Birch, H.G., Grabie, E., and Yoder, K. (1972) The behavior of children in school who were severely malnourished in the first two years of life. *J. Health Soc. Behav.* 13:276–286.

———, ———, and Ragbeer, C. (1975) The behavior of children at home who were severely malnourished in the first two years of life. *J. Biosoc. Sci.* 7:255–267.

Rider, A.A. and Simonson, M. (1974) The relationships between maternal diet, birthweight, and behavior of the offspring in the rat. *Nutr. Rep. Intern.*, 10:19–24.

——— and ——— (1973) Effect on rat offspring of maternal diet deficient in calories but not in protein. *Nutr. Repts. Inter.* 7:361–370.

Riess, B.F. and Block, R.J. (1942) The effect of amino acid deficiency on the behavior of the white rat. I: Lysine and crystine deficiency. *J. Psychol.* 14:101–113.

Riopelle, A.J. (1974) Early protein deprivation and behavioral deficit reconsidered. In: J. Cravioto, L. Hambraeus, and B. Vahlquist (eds.), *Early Malnutrition and Mental Development.* Uppsala, Sweden: Almquist and Wiksell, pp. 127–133.

Robson, J.R.K. (1978) *Malnutrition: Its Causation and Control.* New York: Gordon and Breach.

Rosenn, D.W., Loeb, L.S., and Jura, M. (1980) Differentiation of organic from nonorganic failure to thrive syndrome in infancy. *Ped.* 66:698–704.

Rosenzweig, M.R. and Bennett, E.L. (1972) Cerebral changes in rats exposed individually to an enriched environment. *J. Comp. Physiol. Psych.* 80:304–313.

———, ———, and Diamond, M.C. (1972) Chemical and anatomical plasticity of brain: Replications and extensions. In: J. Gaito (ed.), *Macromolecules and Behavior*, 2nd ed. New York: Appleton-Century-Crofts, pp. 205–278.

Rossetti-Ferreira, M.C. (1978) Malnutrition and mother-infant asynchrony: Slow mental development. *Inter. J. Beh. Dev.* 1:207–219.

Rush, D. (1984) The behavioral consequences of protein-energy deprivation and supplementation in early life: An epidemiological perspective. In: J.R. Galler (ed.), *Nutrition and Behavior.* New York: Plenum, pp. 119–157.

——— and Krystal, A.R. (1982) Methodologic studies during pregnancy: The reliability of the 24-hour dietary recall. *Am. J. Clin. Nutr.* 35:1259–1268.

———, Stein, Z., and Susser, M. (1980) A randomized controlled trial of prenatal nutritional supplementation in New York City. *Pediatrics* 65:683–697.

Rutter, M. (1979) Maternal deprivation, 1972–1978: New findings, new concepts, new approaches. *Child Dev.* 50:283–305.

Scarr, S. (1981) Testing for children: Assessment and the many determinants of intellectual competence. *Am. Psychol.* 36:1159–1166.

Scrimshaw, N.S. (1981) Significance of the interactions of nutrition and infection in children. In: R.M. Suskind (ed.), *Textbook of Pediatric Nutrition.* New York: Raven, pp. 229–240.

——— and Behar, M. (1961) Protein malnutrition in young children. *Science* 133:2039–2047.

——— and Gordon, J.E. (1967) Editor's note. In: N.S. Scrimshaw and J.E. Gordon (eds.), *Malnutrition, Learning and Behavior.* Cambridge, MA: MIT Press, pp. 288–289.

Second National Health and Nutrition Examination Survey: Dietary Intake Source Data: United States, 1976–1980. (1983) U.S. Department of Health and Human Services. (DHHS Pub. No. (PHS)83–1681)

Shaywitz, S.E., Cohen, D.J., and Shaywitz, B.A. (1978) The biochemical basis of minimal brain disfunction. *J. Pediatr.* 92:179–187.

Shoemaker, W.J. and Bloom, F.E. (1977) Effect of undernutrition on brain morphology. In: R.J. Wurtman and J.J. Wurtman (eds.), *Nutrition and the Brain, Vol. 2.* New York: Raven.

Sigman, M., Cohen, S.E., and Forsythe, A.B. (1981) The relation of early infant measures to later development. In: S.L. Friedman and M. Sigman (eds.), *Preterm Birth and Psychological Development.* New York: Academic Press, pp. 313–328.

Simonson, M. (1979) Effects of maternal malnourishment, development, and behavior in successive generations. In: D.A. Levitsky (ed.), *Malnutrition, Environment, and Behavior.* Ithaca, NY: Cornell University Press, pp. 138–148.

———, Stephan, J.K., Hanson, H.M., and Chow, B.F. (1971) Open field studies in offspring of underfed mother rats. *J. Nutr.* 101:331–335.

Singer, L.P. and Fagan, J.F. (1983) Cognitive development in the failure to thrive infant: A three year longitudinal study. *J. Ped. Psychol.* 9:363–383.

Slob, A.K., Snow, C.E., and Mathot, E. (1973) Absence of behavioral deficits following neonatal undernutrition in the rat. *Dev. Psychobiol.* 6:177–186.

Smart, J.L. (1984) Animal models of early malnutrition: Advantages and limitations. In: J. Brozek and B. Schurch (eds.), *Malnutrition and Behavior: Critical Assessment of Key Issues.* Lausannė, Switzerland: Nestle Foundation, pp. 449–459.

——— (1974) Activity and exploratory behavior of adult offspring of undernourished mother rats. *Dev. Psychobiol.* 7:315–321.

——— and Dobbing, J. (1972) Vulnerability of developing brain. IV. Passive avoidance behavior in young rats following maternal undernutrition. *Dev. Psychobiol.* 5:129–135.

———, ———, Adlard, B.P.F., Lynch, A., and Sands, J. (1973) Vulnerability of developing brain. Relative effects of growth restriction during the fetal and suckling period on behavior and brain composition of adult rats. *J. Nutr.* 103:1327–1338.

Smith, D.W. (1977) *Growth and Its Disorders.* Philadelphia: W.B. Saunders Co.

Solomons, N.W. (1985) Assessment of nutritional status: Functional indicators of pediatric nutriture. *Pediatric Clinics of North America* 32:319–334.

Spitz, R.A. (1945) Hospitalism: An inquiry into the genesis of psychiatric conditions of early childhood. *Psychoanal. Study of the Child* 1:53–74.

Srikantia, S.G. (1982) Protein-energy malnutrition. In: A. Neuberger and T.H. Jukes (eds.), *Human Nutrition: Current Issues and Controversies.* Englewood, NJ: Burgess, pp. 197–207.

Sroufe, L.A. and Waters, E. (1977) Attachment as an organizational construct. *Child Dev.* 48:1185–1199.

——— (1979) Socioemotional development. In: J. Osofsky (ed.), *Handbook of Infant Development.* New York: John Wiley, pp. 462–516.

Staub, E. (1978) *Positive Social Behavior and Morality, Vol. 1.* New York: Academic Press.

Stein, D.G. (1974) The effects of early saline injections and pentylenetetrazol on Hebb-Williams maze performance in the adult rat. *Behav. Biol.* 11:415–422.

Stoch, M.B. and Smythe, P.M. (1976) 15-year developmental study on effects of severe undernutrition during infancy on subsequent physical growth and intellectual functioning. *Archs. Dis. in Child.* 51:327–336.

——— and ——— (1967) The effect of undernutrition during infancy on subsequent brain growth and intellectual development. *S. Afr. Med. J.* 41:1027–1030.

——— and ——— (1963) Does undernutrition during infancy inhibit brain growth and subsequent intellectual development? *Arch. Dis. Child* 38:546–552.

———, ———, Moodie, A., and Bradshaw, P. (1982) Psychosocial outcome and CT findings after gross undernourishment during infancy: A 20-year developmental study. *Dev. Med. Child Neurol.* 24:419–436.

Strayer, F.G. and Strayer, J. (1976) An ethological analysis of social agonism and dominance relations among preschool children. *Child Dev.* 47:980–989.

Strobel, D.A. (1972) Stimulus change and attentional variables as factors in the behavioral deficiency of malnourished developing monkeys (Macaca mulatta). Unpublished doctoral disseration, University of Montana.

——— and Zimmerman, R.R. (1972) Responsiveness of protein deficient monkeys to manipulative stimuli. *Dev. Psychobiol.* 5:291–296.

——— and ——— (1971) Manipulatory responsiveness in protein-malnourished monkeys. *Psychonom. Sci.* 24:19–20.

Strupp, B.J. (1982) Malnutrition and animal models of cognitive development. Unpublished doctoral dissertation, Cornell University, Ithaca, NY.

——— and Levitsky, D.A. (1983) Early brain insult and cognition: A comparison of malnutrition and hypothyroidism. *Dev. Psychobiol.* 16:535–549.

Stuart, H.C. and Meredith, H.V. (1946) Use of body measurements in the school health program. *Amer. J. of Public Health* 36:1365–1386.

Sukhatme, P.V. and Margen, S. (1982) Autoregulatory homeostatic nature of energy balance. *Am. J. Clin. Nutr.* 35:355–365.

Suskind, R.M. (1981) Malnutrition and the immune response. In R.M. Suskind (Ed.). *Textbook of Pediatric Nutrition.* New York: Raven.

——— and Varma, R.N. (1984) Assessment of nutritional status of children. *Peds. in Rev.* 5:195–202.

Tanabe, G. (1972) Remediating maze deficiencies by the use of environmental enrichment. *Dev. Psycho.* 7:224.

Tanner, J.M., Goldstein, H., and Whitehouse, R. (1970) Standards for children's height at age 2–9 years allowing for height of parents. *Arch. Dis. Child* 45:755–58.

Ten-State Nutrition Survey in the United States, 1968–1970. (1975) Washington, D.C.: DHEW. (DHEW Publ. No. (HRA) 74–1219)

Tizard, B. and Hodges, J. (1978) The effect of early institutional rearing on the development of eight-year-old children. *J. Child Psychol. and Psych.* 19: 99–118.

——— and Rees, J.A. (1974) A comparison of the effects of adoption, restoration to the natural mother, and continued institutionalization on the cognitive development of four-year-old children. *Child Dev.* 45:92–99.

Trowbridge, F.L. (1984) Prevalence of growth stunting and obesity: Pediatric nutrition surveillance system. *Mortality and Morbidity Weekly Report.* 32: 23–26ss.

Trowell, H.C., Davies, J.P.N., and Dean, R.F.A. (1954) *Kwashiorkor in Children.* London: Arnold.

Tuddenham, R.D. (1974) On analyzing longitudinal data relating nutrition to cognitive development. In: J. Cravioto, L. Hambraeus, and B. Vahlquist (eds.), *Methodology in Studies of Early Malnutrition and Mental Development.* Uppsala, Sweden: Almqvist and Wiksell, pp. 72–82.

Underwood, B.A., et al. (1967) Height, weight and skinfold thickness data collected during a survey of rural and urban populations of West Pakistan. *Am. J. Clin. Nutr.* 20:694–701.

Usher, R. and McLean, F. (1969) Intrauterine growth of live-born Caucasian infants at sea level: Standards obtained from measurements in 7 dimensions of infants born between 25 and 44 weeks of gestation. *J. Ped.* 74:901–910.

Vahvelainen, M.L. and Oja, S.S. (1972) Kinetics of influx of phenylalanine, tyrosine, tryptophan, histadine, and leucine into slices of brain cortex from adult and 7-day-old rats. *Brain Res.* 40:477–488.

Valman, H.B. (1974) Intelligence after malnutrition caused by neonatal resection of ilium. *Lancet* 1:425–427.

Vietze, P.M., Falsey, S., O'Connor, S., Sandley, H., Sherrod, K., and Altemeier, W. (1980) Newborn behavioral and interactional characteristics of nonorganic failure to thrive infants. In: T. Field (ed.), *High Risk Infants and Children: Adult and Peer Interaction*. New York: Academic Press, pp. 5–23.

Villar, J., Smeriglio, V., Martorell, R., Brown, C.H. and Klein, R.E. (1984) Heterogenous growth and mental development of intrauterine growth-retarded infants during the first three years of life. *Pediatrics* 74:783–791.

Viteri, F.E. (1981) Primary protein-energy malnutrition: Clinical, biochemical, and metabolic changes. In: R.M. Suskind (ed.), *Textbook of Pediatric Nutrition*. New York: Raven, pp. 189–216.

Vore, D.A. and Ottinger, D.R. (1970) Maternal food restriction: Effect on offspring development, learning, and a program of therapy. *Dev. Psychobiol.* 3:337–342.

Vuori, L., Christiansen, N., Clement, J., Mora, J.O., Wagner, M., and Herrera, M.G. (1979) Nutritional supplementation and the outcome of pregnancy. II. Visual habituation at 15 days . *Am. J. Clin. Nutr.* 32:463–469.

Waber, D.P., Vuori-Christiansen, L.V., Ortiz, N., Clement, J.R., Christiansen, N.E., Mora, J.O., Reed, R.B., and Herrera, M.G. (1981) Nutritional supplementation, maternal education, and cognitive development of infants at risk of malnutrition. *Am. J. Clin. Nutr.* 34:807–813.

Walsh, R.N. and Cummins, R.A. (1975) Mechanisms mediating the production of environmentally induced brain changes. *Psychol. Bull.* 82:986–1000.

——— and ——— (1976) Neural responses to therapeutic environments. In: R.N. Walsh and W.T. Greenough (eds.), *Environments as Therapy for Brain Dysfunction*. London: Plenum, pp. 171–200.

Warren, N. (1973) Malnutrition and mental development. *Psychol. Bull.* 80:324–328.

Waterlow, J.C. (1984) Current issues in nutritional assessment by anthropometry. In: J. Brozek and B. Schurch (eds.), *Malnutrition and Behavior: A Critical Assessment of Key Issues*. Lausanne, Switzerland: Nestle Foundation, pp. 77–90.

——— (1974) Some aspects of childhood malnutrition as a public health problem. *Brit. Med. J.* 4:88–90.

——— (1972) Classification and definition of protein-calorie malnutrition. *Brit. Med. J.* 3:566.

———, Cravioto, J., and Stephen, J.M.L. (1960) Protein malnutrition in man. In: C.B. Anfinson et al. (eds.), *Advances in Protein Chemistry*. London: Academic Press, p. 131.

Waters, E. (1979) The reliability and stability of individual differences in infant-mother attachment. *Child Dev.* 49:483–494.

Wegman, M.E. (1985) Annual summary of vital statistics–1984. *Pediatrics* 76: 861–871.

Wellcome Working Party (1970) Classification of infantile malnutrition. *Lancet* ii:302–303.

Whitten, C.F., Pettit, M.G., and Fischhoff, J. (1969) Evidence that growth failure from maternal deprivation is secondary to undereating. *J. Amer. Med. Assoc.* 209:1675–1681.

Wiener, S.G., Robinson, L., and Levine, S. (1983) Influence of perinatal malnutrition on adult physiological and behavioral reactivity in rats. *Physio. and Beh.* 30:41–50.

Winick, M. (1976) *Malnutrition and Brain Development.* New York: Oxford University Press.

———, Brasel, J.A., and Rosso, P. (1972) Some speculations on mechanisms involved in the effects of undernutrition on cellular growth. In: *Nutrition, the Nervous System, and Behavior.* Pan American Health Organization, Scientific Publ. No. 25, pp. 24–32.

——— and Morgan, B. (1985) Nutrition and brain development. In: W.A. Walker and J.B. Watkins (eds.), *Nutrition in Pediatrics: Basic Science and Clinical Application.* Boston: Little, Brown and Company, pp. 233–252.

——— and Noble, A. (1966) Cellular response in rats during malnutrition at various ages. *J. Nutr.* 89:300–306.

World Health Organization (1983) *Measuring Change in Nutritional Status: Guidelines for Assessing the Nutritional Impact of Supplementary Feeding Programmes for Vulnerable Groups.* Geneva: World Health Organization.

——— (1978) *A Growth Chart for International Use in Maternal and Child Health Care: Guidelines for Primary Care Personnel.* Geneva: World Health Organization.

Wretlind, A. (1982) Standards for nutritional adequacy of the diet: European and WHO/FAO viewpoints. *Am. J. Clin. Nutr.* 36:366–375.

Wurtman, R.J., and Fernstrom, J.D. (1976) Control of brain neurotransmitter synthesis by precursor availability and nutritional state. *Biochem. Pharmacol.* 25:1691–1696.

———, Hefti, F., and Melamed, E. (1980) Precursor control of neurotransmitter synthesis. *Pharmacol. Rev.* 32:315–335.

Wyman, A. (1984) Is learning impaired by hunger? *The Boston Globe*, February 20, pp. 37,39.

Yaktin, U.S. and McLaren, D.S. (1970) The behavioral development of infants recovering from severe malnutrition. *J. Ment. Defic. Res.* 14:25–32.

Yarrow, M.R., Campbell, J.D., and Burton, R.V. (1970) Recollections of childhood: A study of the retrospective method. *Mono. Soc. Res. in Child Dev.* 35 (Whole no. 138).

———, Scott, P.M., and Waxler, C.Z. (1973) Learning concern for others. *Dev. Psychol.* 8:240–260.

———, and Waxler, C.Z. (1979) Observing interaction: A confrontation with methodology. In: B. Cairns (ed.), *Social Interactional Analysis: Methods and Illustrations.* Hillsdale, NJ: Lawrence Erlbaum, pp. 37–65.

Zerfas, A.J., Shorr, I.J., and Neumann, C.G. (1977) Office assessment of nutritional status. In: C.G. Neumann and D.B. Jelliffe (eds.), *The Pediatric Clinics of North America* 24:1, Symposium on Nutrition in Pediatrics. Philadelphia: W.B. Saunders Company, pp. 253–272.

Zeskind, P.S., Goff, D.M., and Huntington, D.L. (1984) A developmental approach to the study of the behavioral effects of early malnutrition. In: J. Brozek and B. Schurch (eds.), *Malnutrition and Behavior: Critical Assessment of Key Issues.* Lausanne, Switzerland: Nestle Foundation, pp. 424–440.

——— and Lester, B.M. (1978) Acoustic features and auditory perceptions of the cries of newborns with prenatal and perinatal complications. *Child Dev.* 48:580–589.

——— and Ramey, C.T. (1981) Preventing intellectual and interactional sequelae of fetal malnutrition: A longitudinal, transactional, and synergistic approach to development. *Child Dev.* 52:213–218.

——— and ——— (1978) Fetal malnutrition: An experimental study of its consequences on infant development in two caregiving environments. *Child Dev.* 49:1155–1162.

Zigler, E. and Butterfield, E.C. (1968) Motivational aspects of changes in IQ performance of culturally deprived nursery school children. *Child Dev.* 39:1–14.

——— and Trickett, P.K. (1978) IQ, social competence, and evaluation of early intervention programs. *Am. Psychol.* 33:789–798.

Zimmerman, R.R., Geist, C.R., Strobel, D.A., and Cleveland, T.J. (1974) Attention deficiencies in malnourished monkeys. In: J. Cravioto, L. Hambraeus, and B. Vahlquist (eds.), *Early Malnutrition and Mental Development.* Uppsala, Sweden: Almquist and Wiksell, pp. 115–126.

———, ———, and Wise, L.A. (1974) Behavioral development, environmental deprivation, and malnutrition. *Advs. in Psychobiol.* 2:133–191.

———, Steere, P.O., Strobel, D.A., and Hom, H.L. (1972) Abnormal social development of protein-malnourished rhesus monkeys. *J. Ab. Psych.* 80: 125–131.

———, Strobel, D.A. (1969) Effects of protein-calorie malnutrition on visual curiosity, manipulation, and social behavior in the infant rhesus monkey. *Proc. 77th Annual Conv. APA* 5:241–242.

———, ———, and Maguire, D. (1970) Neophobic reactions in protein-malnourished infant monkeys. *Proc. 78th Annual Conv. APA* 6:197–198.

——— and Wells, A.M. (1971) Performance of malnourished rats on the Hebb-Williams closed field maze learning task. *Per. Motor Skills* 33:1043–1050.

AUTHOR INDEX

Subject Index